The R.A.M.S. Library of Alchemy

Volume 36

The Golden Fleece

by

Salomon Trismosin

Also includes:

OF NATURE AND ART

LIBER TRIUM VERBORUM OF KING CALID

THE PHILOSOPHICAL CANONS OF PARACELSUS

R.A.M.S. Publishing Company

The Golden Fleece

by

Salomon Trismosin

Also includes:

OF NATURE AND ART

LIBER TRIUM VERBORUM OF KING CALID

THE PHILOSOPHICAL CANONS OF PARACELSUS

Produced by

Restorers of Alchemical Manuscripts Society
1982

R.A.M.S. Publishing Company

R.A.M.S. Publishing Company
117 Rutherford Lane
Stuarts Draft VA 24477

The Golden Fleece

First Edition 2015

ISBN-13 **978-1511934879**
ISBN-10 **1511934875**

Image Processing by Philip N. Wheeler

Printed in the United States of America

Table of Contents

Dedicated to Hans W. Nintzel,

American Alchemist

and

Founder of the

Restorers of Alchemical Manuscripts Society

(R.A.M.S.)

THE GOLDEN FLEECE

OR

THE FLOWER OF TREASURES

In which is succinctly and methodically
handled, the Stone of the philosophers,
his excellent effect *&* admirable
Virtues: and
The better to attain to the Original
& true means of Perfection.

Enriched with Figures representing
the Colours to rise as they
successively appear in the
Practice of this
Blessed Work.

By that great Philosopher
SALOMON   TRISMOSIN
Master to Paracelsus.

Disclaimer

Liability: The publisher does not warrant or assume any legal liability or responsibility for the accuracy, completeness, or usefulness of any information, apparatus, product, or process disclosed. The publisher makes no representation as to the accuracy or completeness of the contents of this book and specifically disclaims any implied warranty of merchantability or fitness for a particular purpose. No warranty may be created or extended by written sales materials or sales representatives. You should obtain professional consultation where appropriate. The publisher shall not be liable for any loss of profit or other commercial or personal damages, including but not limited to special, incidental, consequential, or other damages.

Introduction

Philip N. Wheeler

Salomon Trismosin was an adept and teacher of Paracelsus. He also wrote Splendor Solis, a well-known text with 22 color plates. Trismosin was born in Germany in the 15th century. His "Golden Fleece" was almost as popular as "The Hieroglyphics and Testament of Nicolas Flamel."

Descriptions of various enigmatic figures are included throughout "Golden Fleece." Unfortunately, Hans Nintzel did not reproduce the illustrations themselves.

This Volume also includes the works originally presented in the R.A.M.S. release, "Three Tracts":

OF NATURE AND ART

LIBER TRIUM VERBORUM OF KING CALID

THE PHILOSOPHICAL CANONS OF PARACELSUS

PREFACE

ALPHIDIUS, (truly esteemed one of the most famous & the most recommended to posterity; among the Ancient & sage Philosophers of his time) do propose to us in his divine writings; that the ordinary Contemplation, Mysterious consideration, and the continual reading of approved and renowned, chiefly contended Authors, and some who have more dismally delivered this work to us, wonderfully and never sufficiently praised, admired and reverenced of the most oppollent spirits; who have followed the pursuit through Curiosity or for compassion, seeing so many poor souls, consuming their days in Illusions, have thought it convenient to bring to light some glittering tinsel of that most oppollent work of our Lyon which is known by his paws in earnest of the most Spiritual light which they have contrived it being come somewhat near to this precious Stone, one might be approved by this sacred small spark.

This wise and sage Doctor saith that the Inquisitors of this terrestrial Son shall receive as much or more fruit and contentment of the Learned Nutrients sprang from the provident Tutelage of this more than desired and without doubt Celestial Stream of amiable food agreeable to the delightful and sweet breast-milk, then he doth discontent and mistake in the dull discontent and mistake in the dull sound of the Ignorant sorts who are not sufficiently enabled to Judge discreetly and to apprehend the depth of so profound, grand and Occult a Mystery; their Light not being subtil enough to perceive this Subject and their Brains not Judicious enough to prize this inestimable Pearl.

But rather only nourished, elevated, comforted and satisfied with vain hope or to speak better they being held back by Bitter Juice of Ignorance, are made incapable of the more Solid food for to direct them and to return them again to their proposed (as a mark set before their eyes) Art of the Stone of the Sages the which we call the Heaven of the Philosophers: But I do not write unto those, but rather that they do not immerse themselves in the intricate ambiguity of the Golden Fleece nor not at all to touch the Sauce with the least point of their fingers, nor that they assaile the inestimable Labourenth with their weak abilities, because these heady brainless are not at all rallied to the glorious triumph of this degree; of him only given to those that love Wisdom, by no means to all busy-brains, taken with a fantastical conceit attempting to snake the delicious honey of our judicious writings.

It were better and more fittable for these dulipates to have considered it: Lost thy charge before they began their Labour not to have anything, how false should it be from our Divine Work but rather to have retired from the fleeting Garden of the delightfull Hesperides.

The senseless noise of their Inabilities incapable of the propositions (too subtil for their pates) of our oppollent Work in regard of the disposition of their feeble fancy's.

This Celestial Gift doth not at all amaze us in the General Canopy of the whole world in gross; but in the retail, considering and ought to be dispised, and especially favouring and making heros of others as of those that may be known to be the true Sons of this Science, calling them the Blessed of the most happy Rays of the Golden Branches from the which the others are driven away, as from their Sacred Fires.

Profane approach not this Treasure Sacred:

For Holy Ones only Consecrated.

RASIS saith not less in the Treatise he made of the Light of Lights; now ought saith he, so to presume of him as also to run so that assured hope lest he be plainly blameable for his evil desert for stretching his desires beyond the Imprudent Limits of his Capacity following his own will in the feeble Authority of his weak Spirit the pure and clean Essence of the admirable mixtures not withstanding they have not known the perfect Elements. But to speak truth, these sort of People bray of more than they have gotten, they have principally more confusion then contention, more sorrow than Solace; a thousand times more cause to be reprehended with a sharp Chastisement, then to have fruit-full gain of their purpose which call to mind Apelles his Quip, which he gave in two sentences touching the presumption of an Arrogant fellow steadily

chastiseing at that Instant when he did reprehend
the malacious discourse of a simple Cobbler, to
redeem the venerable censure of the line and
portraiture of his oppollent picture.

Thou maist speak saith he of the pantable:

But if thy doubtest thou art not able.

Also very well to this purpose the better to avoid
the evil speech and censure of a publick obligation,
he hath set before us this point on modesty.

Attempt not more than thou art able;

Who boast yet knows not doth but babble.

On the other Column which he set to prepare and
sustain it.

Exercise thy knowledge of thy Art

Beyond Experience do not start.

But nowadays very many do set miserably, flourish
and flattering, and infatuating themselves with the
vain hope of their own opinions when they have found
nought more than a dole of which boldly they take in
hand not one penny, they thinking in this Iron age

to get sheckles of Gold more surely than the Bean (acorn?) in the Oak.

Gold the Alchymist so long sought

Till at length he is brought nought.

In this wise, phantastically supposing they have now got the moisture of the moon Breaking their Brains with the concept how to make the Moon with her influence descend upon the body of the earth the Mother of the Elements by a way which notwithstanding they have not known, only being supported by natural appearances and covetous of Curiosities and desirous of Novelties. But if it be true, IGNOTI NULLA CUPIDO, according to Philosophy a ripe Ingenuity may conceive by applyance of transcended efforts.

Their spirits more slight than Clouds blown;

Cannot speak truth of a thing unknown.

And no more than blind men deprived of their Sight can Judge of colours. No more can these Ignorant fellows speak of the Heaven of the Philosophers, then their feet under the table.

Si te fata vocant aliter non
Saith AUGURELLIUS in his Chrysopeia.

They only whom the Heavens favours

With this precious Art are blest

Tis ordained for no others:

All but wise Sages have it missed.

Also I shall enlarge myself further for their better
apprehension, so they will study to unfold the
Intracateness of this business which is not easily
done by the Importunity of the rash practioners of
this Science. All those who have imployed and
presented their Bark to the gulf have not all
arrived at the Haven, yea, the most of those who
have imbarked themselves for this Port have by a
thousand mishaps suffered shipwreck.

The Wise Argonauts conducted in the Waves by the
insistant hand of a long destiny, after a thousand
Crosses; in the end conquered this rich Fleece by
their valour armed and supported by the industry of
Experience and patience, the true Conduct to a quiet
Haven necessary in this Work.

Paucy quos aquus amavit.

Jupiter aut ardens evescit ad aethera virtus.

God gives not this especial favour

But to them of heavens Graces savour.

Also he ought to arrive at the famous Ille without calling Colchos, the better to prevent shipwreck and to come to the points of the natural Causes one should have all his fingers and the best of the Ancient Philosophers and to unfold their Writings and to Judge of the verity by the Concordance of their several descriptions, otherwise they have lost the false guide of the Intricate Labourinth only in their books hiding it from the ignorant.

Dare ye take with Sacrilegious hand

Our chief Secret without our command.

No No retire ye have not such slight

To ensnare the bird of our delights.

The Philosophers are cautious to talk or discourse, but with their like, yea, they do not speak but to the most knowing man as it is said in the Complaint of Nature.

Whats need to thee I shew if thou art Sage

If not, not by me hast thou no advantage.

This is the cause why they command that their books should by no means contain anything that might make dull and ignorant people capable to obtain the sweet honey of so many flowers.

ROSINUS conformable to former Authors doth not
approve that in any wise, men of weak Spirits,
should employ their Wits in this Assay without the
knowledge of that, the Philosophers have not named
in their writings. Where is concord there is verity
saith the Count Trevisan in his great Rosary
Concorda Philosophos and bene rily erit.

If seeming discord thou canst make concord

And concord Sages, accord some discord.

The which one ought to enter prize by this Art and
principal natural not familiar, but by a secret
ground and more clear than day, the corporeal things
take their substance and essence of the terrestrial
mass, for the earth is the mother of the Elements
from the earth they proceed and to earth they return
saith HERMES.

The Element of the Earth is General

Mother, and in her womb he doth nourish all.

As it were the Vessel of generation because their
Philosophers accord to the order of the time, of the
influence of Heaven (doth serveth for it for seed
and fermentative heat to make it spring and bring
forth matter) of the Planets, of the Sun, of the
Moon, or of the Stars and so of others consequently
with the 4 qualities of the Elements, which serving

one and the other for wombs, they move without ceasing.

And by this means all things fruitfully increasing and growing by a form and original peculiar to their proper substances according to the Almighty power and divine Will which maketh things as at the first moment of the Admirable Creation of the World.

The metals also, are according to the Course of other Created things taking their Original from the Earth, Mother of the Elements and Nurse of all things as I have afore declared with a matter proper and Individually derived full and wholly from the four properties of the Elements by the Influence agreeing to the power of the metals and the Conjunction of the Constellations of the Planets. ARISTOTLE in the 4th of the Meteors is of the same opinion, where he saith and affirmeth that the Argent Vive is, the true matter Common to all the metals. But now nature Earth first gathered and conjoined together a matter of the 4 Elements only thereof composing a substance according to the effort and propertie of the matter which the Philosophers name MERCURY or ARGENT VIVE NOT COMMON or made by Art, but rather having a form perfected by Gold and Silver, or drawn from the Imperfect metals. The curious naturalists concerning the nature of metals have spoken clearly enough in their Books and therefore there is no need here to be over Long unless it be upon this assured solid Base, the proper ground principal, and Mastery of the Stone of the Sages. The Original of which is found in the Center and perfect Body of Nature which is not taken

from anything living and of the same only thing we
have the means to obtain the perfect formed, and the
most great Contentment of the final Perfection.

THE GOLDEN FLEECE

Of the Original of the

Stone of the Sages & how

it may be brought to

Perfection.

This Stone of Wise men draws the pure Elements of his Essence by the assured way of a mindful and fundamental nature in the which also she amends herself as HALY reports when he saith of this Stone; "It makes his own influction and Imbibition of things growing and secretly conjoining themselves, congealing and resolving itself by the nature which bettereth the things and maketh it more perfect and of greater efficacy, orderly and according to the time ordained." Upon this model and such like pattern of Artifice must every man rely himself and inform himself to these natural principles. If he desires to have help and assistance in his Arts by the operation of nature which in amenities it preforms itself until the time comes that by his natural Art he perfects the true form of his Intention.

Now this Artifice is no other thing than one only operation and perfect preparation of the matter, which wise and prudent nature hath made, in the Moisture of this Foundation. To this which also agrees a Moderiety of proportion and an assured measure of this operation with a mature foundation and considerate prudence for although Art may take

of itself the Sun and Moon for a new beginning, as
it were to make Gold yet there is a necessity of
knowing the natural Secret of matters Mineral and
how in the Entrals of the earth they have the
foundation of their first principals, it being most
certain the Art affords another way than Nature
having to this Effect another and altogether diverse
operation. It is also convenient to this Artifice
producing out of those preceding natural roots in
the beginning of nature should produce exquisite
things, which nature of herself could not have
operated. For true it is that she hath not in her
power the ability to engender those things of
herself by the which natural metals come to
procreate themselves in a long time almost
altogether imperfect, notwithstanding incontinendly
after and almost in a moment may be made perfect by
the rare secret of the ingeneous Artist, this which
proceeds from the temporal matter of nature. And
which serves to this Artifice of men, when she
comforts them with her free means and then again Art
aids her by his timely operation, but in such sort
that this form being complete, may afterwards hold
correspondence and make itself fit to the first
Intention of nature and last perfection of his
design. And although with great Artifice this Stone
above-mentioned should return to the proportion of
his first form, that being whereof it draws from the
treasure of nature (as all other the substantial
form of things grow into diverse fashions like
Animals, or metals) yet do they all proceed of one
inward power of the matter; only excepted the Soul
of man which is not so limited, nor rely as (as
those other things under this terrestrial and
temporal submission) But take heed also and it is
very considerable that the substantial form hath no

relation nor can condescend to the matter, were it not that it is done by a certain operation of some accidental form; Nor yet that this happens by his own peculiar force, but rather by means of some other operation sustained as the fire or some such heat as nearly correspondent thereunto perfectly joined that must work thereupon wherein the better to oppress ourselves and to render our position more Intelligable. We will take the Similitude of a Hens Egg. In the which subsists the substantial form, putrefaction, without the Accidental form, that is to say a mixture of red and white by the particular power of an Internal and natural heat working therein. But although the Egg be the matter of the Chicken the form notwithstanding is neither substantually nor accidently comprized therein but only potentially. For putrefaction which is the principal of all generation is indured by this means and with the assistance of an outward heat according to the Maxime which says: Heat working in moisture bringeth first blackness and after working upon dryness promotes whiteness.

Even so it is in the natural matter of the Stone in the which consists neither the Substantial, nor accidental form without putrefaction or decoction which brings it to be in power, which it is afterwards in effect, now it rests to demonstrate and shew of so habitude this Putrefaction is that is so necessary to generation and from whom it hath its original. Corruption or putrefaction are sometimes engendered by an optimum heat it being put in some certainly hot place, or by the warmth which is drawn from something that hath moisture in it, this putrefaction is likewise made of a superfluous coldness when natural heat decays and is dispersed weakening and corrupting by an over-abounding coldness which is properly called privation where natural heat abandons the thing and such a Corruption is assuredly made in things that are cold and moist. Philosophers speak not at all of this kind of putrefaction, but of corruption which is no other thing then humidity in Siccity by the means whereof all dry things come to be dissolved, joining the fire with the water as TREVISAN saith to return to resume their first being upon which they pretend soon after according to the property of their nature to arrive to the preference of their final form.

In this Corruption the moistness itself, with the dry (which notwithstanding is not so dry) but by the moisture is kept by intermingling itself with that which is dry. And therefore it may more properly be made an Impression of Spirits or a Plain Congealation of matters. But when the moist comes to disunite itself and make an entire Separation from the dry we must instantly withdraw the dry part and reduce it into Ashes. So the Philosophers intend that their corruption, dryness, disruption or dissolution, and Calcination be done in such sort as the natural moisture and dryness do rejoice themselves, dissolve and reunite together by an abundance of moisture and dryness and by an equal proportion of temperature to the End that with great facility superfluous and corruptible things may be drawn and vapoured away as unprofitable and sootie Excrements.

Neither more nor less then as the meat taken into the Stomach, assimilates properly and conducts itself into the Substance of the nature it is nourished when it is seasoned by a digestion and laudable concoction, and when by the proportion, and digestion made in the Ventricle she draws unto herself a Plain substantial and convenient humidity.

Now by the means of this radical humour, nature is conserved and augmented, their sootie and superfluous parts over abounding, as a corrupted Sulphur being rejected from them. But it is remarkable that anyone of their parts will be nourished according to the property of his nature, in the which he rejoyceth and desires there to remain and to conserve his individual in the same Species which we ought also as well to understand of this Stone of wise men as of the human body which changes into the purity of his substance the inferiour form though of a different condition by the means of this natural and well-tempered fire which is the true governour and the only guide of our great Mystery as it is said; Minor Ignus omnia tent. The less fire grinds all things. This radical moisture is the Pilot that orders all those diverse natures to live peaceably together and of various contrarie Qualities, and of differing discords compasses an excellent harmony of Agreement by the Industry of a necessary concoction and a moist heat which doth actuate an equal proportion upon their metalline Bodys.

The Body alters all into his proper kind

On what so ere it feeds it nourishment doth find.

So doth our Artifice imperfect metals raise

As they are equalled unto Gold and stand as
strongest Assayes.

The Second Treatise

Representing the Work of the Philosophers

by two Figures.

Here was placed the Figure:

    A Sun in a Shield with a mantle, Helment and Crown upon it:

    The Crest was likewise a Sun.

It is necessary, saith MORIEN, that our operation and the Art whereof we desire at this present to intreate, are divived into 2 principal doctrines, the extremes and the means whereof are strictly tyed together so adhering the one to the other and with such a reciprocal interchange to the immediate and of the first link itself with an indivisible chain to the beginning of the last and do mutually surround one another the cast being lovingly provoked to the Invitation of the selfsame actions the which may be observed and seriously considered in the precedent pattern of her that hath gone before and then is the mastery entirely done and perfected. But they cannot possibly accommodate themselves in any other Body then their own proper matter.

Now the better to conceive and understand more assurdly, it is necessary in the first place to note the Nature according to GEBER assines out of the first Essence of metals composed of Mercury and

Sulphur which opinion is seconded by the Authority
of FERRARIUS in his Questions of Alchemy, 25th.
Chapter, where he saith that nature proceeds from
the Original and Pure Essence of metallick nature
the which in the fire becomes a putrefying water
which she mingles with a Stone very white and
subtile rendering and resolving it as it were into a
broth and certain Vapour raised out of the veins of
the Earth, the which she doth churn by force of a
continual motion to make her digestion, and doth
vapour together with equal moisture and dryness
uniting and coagulating themselves in such sort as
from them is produced a slimy Substance which we
commonly call MERCURY or QUICKSILVER:

The which is no other thing then the Spring-head or
first matter of Metals as we have heretofore
affirmed. And further the same Author testifies to
us in his 26th. Chapter, that those that will set as
far as it is lawful and possible to follow nature
must not only help themselves with Quicksilver only,
but with Quicksilver and Sulphur both together, the
which notwithstanding they must not only mingle but
also especially prepared, and season with Wisdom the
which prudent nature hath produced and reduced into
a perpetual confluxure.

Now so it is the which such a kind of Argent Vive
nature begins her first operation and finisheth by
the way of metals with which she contents herself
for the entire perfection of her work she having
thus accomplished that which belonged to her Duty
and left the rest to Art to accomplish the Intention
in perfecting the Philosophers Stone, and the

absolute framing and forming it to his last period and incomparable Luster, whereby it is evident that we begin the Work where nature hath placed her end and the last glory of her ambition. All the Philosophers agreeing in this Tenent that the true principal of their operation is taken from natures last End, which is the Sun of metals and freely do confess that he that pretends anything in the knowledge of this business or desires perfectly to reach the height of this natural Art must absolutely and without doubt begin at the End and Cessation of Nature and where she reposed herself infine having arrived at the period of her Intension contenting herself with the Achievement of those ordinary operations.

We must then take the Sulphur and the Quicksilver which nature hath brought to a most neat and pure form being accomplished and endowed with a most subtill Union as none other could so ingeniously prepared by no Artifice seen; although the nature (as it is said) finally possesses this matter by the formal Generation of Metals.

Now this matter thus informed by nature doth conduct the Work to the period of perfection: And the Artifice by this means attaineth the safe Port of his design through the force she whereunto being fitly Imbibed and associated to the same matter; To the which Alchymists add Sol to make him dissolve, distinguish his Elements till such time as he hath gotten a nature subtill and spiritual by the purity of Quicksilver and of the nature of Sulphur in which sort that it is become the next approaching matter thereby enabling itself to retain the pure forms of this hidden Stone which matter we call the MERCURY of the Philosophers seeing the two above said conjoined and strictly allied each to the other. The opinion of ARISTOTLE repugnes not this but holds conformity as appears by the advice he gives to King Alexander the Great; Will you (saith he to him) add Gold to the other precious things wherewith Kings are commonly adorned, richly Crowned to the Work of our Stone. I advise you that this MERCURY is alone the matter and the only thing to accomplish our Science notwithstanding it be enfolded in many diversities and ambiguities that very few can assure themselves to find a safe Conduct from the King to approach the Center of this Intricate Labyrinth without the thread of the fair and favourable Ariadine.

Now this obscure diversity shadowed with a hundred ambiguities, paths and veiled with an infinite number of thick dark Clouds is a true touch of the hand of the Philosophers of purpose to mistake it wisely from the Eyes of the Vulgar.

So speaks ROSINUS and EARLE TREVSAN, and all the rest with one Voice to this end that every man by the facility of the work should not indifferently reach to so high a degree lest thereby he despise so precious and inestimable a Jewel having so easily without difficulty gotten and attained the happy period of our work; the perfection of all works, which therefore we call a Collection as being a multitude of things compounded together and a full representation of all the kinds comprehended in nature.

Therefore the Philosophers spake so darkly (sublime the Inferiour by conjoining and distilling, make it again ascend and descend, drying it without and within) with infinite other sentences integrated with such ambiguous and Hyperbolicial figures which notwithstanding we must plienlerly follow and absolutely accomplish if we desire to reap the Notarian fruits of our golden Earth.

Although it seems that ALPHIDIUS doth in some sort oppose himself to this in these terms (we must know, saith he; that when we dissolve and congeal, we sublime also, an Alchymist without intermission of time, do by this means conjoin and purify our Work) and more clearly yet in this which follows, when the Body hath cast into Water and when it begins to calcine it, then all incontinently corrupted becomes black shadowed and obscured; after this hath vanished, he shall be like Lime which sublimed and exalts itself being thus sublimed and dissolved with the Spirit, he putrifies himself; which is a principal or original most worthy to be compared to all the things of this Universe which have life, Soul or Spirit or none be they Mineral, animal or Vegetable Elements or their composition, things cold or hot, and briefly which good things may be found, yea, in earth or heaven is contained or may be apprehended in the possibility of our Art.

These two Doctrines above mentioned signify (according to Philosophers) this black woman and obscure which serves as a Key to all the work which must bear that these rules in our Stone that is to say in blackness, the assured foundation of all the buildings; or else this man which is the form of our matter, which we compare very properly to the Sun.

This may well suffice for the Introduction of the first Document of this Art.

The Second Figure.

A high rocky mountain, which 2 men are breaking the earth with Pick-axes, at the bottom of which a River runs, with the Moon in it; and after , at a bridge, branching itself 2 ways.

Here is declared by many similitudes figurative speeches and philosophical interpretations how we must proceed to the final perfection of the Work.

The Third Figure.

An old Philosopher holding in his hand a Vessel half full of the matter.

The Third Treatise

The great Genius of our Science and Father of the most high and rare Philosophy HERMES in a rapture and entertainment of his Spirit upon the operation of this philosophical Work breaks out into these speeches (this may be termed as it were an end of the world) forasmuch as the Heavens and the Earth seem to bring forth but no man can by this Heaven and this Earth, understand our two precedent Doctrines veiled with so many Hyroglyphicks; many who have attempted this Labour have sweat much before they could attain this perfection, which when once they have gotten they oppress with such Amphibological ambiguities and so confused as they cannot be understood by their figurative and shadowed similitudes being too obscure for those that think to trace their steps and are desirous to embrace the same fortune and be crowned with the same palm seeing they have run (as they think) the same race.

The first Similitude demonstrates unto us that God by his great power and infinite bounty hath created this earth all equal, fat and fertile without sands, without Stones, without mountains or Valleys, by the influence of the Stars and operation of nature, notwithstanding we now see that it retains nothing of the ancient Luster, but rather disfigured from his perfection, that hardly can it be known to be the same thing it was being outwardly changed into diverse forms and figures of strong Stone, of high mountains, and deep Valleys, and inwardly into wonderful things, diverse colours, diverse minerals, diverse salts and sundry metals and although those confused and contrary things are found at this present in the bowels of the Earth, yet proceed they all interiorly from the same first form, then when of a most Large, gross profound, and largeness that it had at the beginning it is brought into a grand and vast Lustre by help of the continual operation of the Sun the heat thereof being always there conserved Vehement, burning and vapourous, mingling itself confusedly even to the very Center of this gross mass with the cold and moist which is shut up in the same body from whence sometimes arise cold Vapours cloudy and aierous, which some of the mixture of those 2 contrary regients, from which being stayed, and inclosed within the Earth in length of time many other vapours do brood so strong that at the last that she is constrained to make way for them to exhale by the opening of the womb, giving unto them (in spite of her power) free passage whereas she would rather have desired to retain them in the natural dens of her most profound Caves where many of them being gathered pell-mell together in long continuance of time raiseth suddenly upon heapes many parts of the earth in one place by the united force of the Exhalations and many others in other places. Yet notwithstanding the mountains and the valleys are to some purpose principally it helpeth the earth to a better temperature of the 4 qualities, heat, cold, moist and dry. Decoction in a manner decocted and diminished. Now in those places we there find the best and purest metals. This reason doth mush inforce that in low grounds where the earth is flat and plain there is not so great quantity of Vapours nor so many Sulphureous exhalations therefore it is more calm and quiet. That which is fat and slimy and where the humidity from above drains itself

downward, and enters thereinto becomes more tender
and soft changing itself into an extreme Whiteness
by the means principally of a drought proceeding
from the heat of the Sun which makes it more strong
more digested and more hardened after a long time.
But a corruptible, frangible sandy earth and which
yet being somewhat tender hanging in Goblets as
Grapes of a Vine is ordinary more wane and by
consequence having less nourishment to compact the
substance thereof is not so lively having retained
little humidity or vigorous nourishment. So it
becoming more difficult to be digested, it being
dissolved in the form of Rowls or other ill-composed
matter.

Now this earth cannot easily be decocted into Stone
if it be not extremely vapoured and replenished with
much humidity. But it is very necessary that with
the drying of the water (which comes of the Vehement
heat and continual burnings of the Sun). The earth
be kept always humid, otherwise this earth should
remain brittle and corruptible and would easily fall
in pieces.

That which notwithstanding hath not yet been
hardened, fully and perfectly may at length be
brought and decocted into a hard and strong Stone by
the continual operation of nature assisted with the
heat of the Sun and a long and continual decoction
without intermission. So the fumes and vapours
aforesaid Shut up in the pores of the earth and when
they come to form with the watery vapours with the
substance of some very subtill earth digested and
well purified by the heat and influence of the Sun,

of the other planets and of all the Elements together; thus may one address and bring to Work QUICKSILVER.

But for as much as it may contract some hardness by a subtill Inflammation one may well serve himself of the Sulphur of the Philosophers and the force and vigour. Thereof the great HERMES concludes very well when he saith, that it shall receive the virtues of the superior and inferior planets and that his force surpasseth and penetrates all other forms, yea even to the very precious Stone.

The Fourth Figure.

A Tree with Birds in it, to which a young man is climbing by a Ladder: The bottom of the Tree grows out of a Crown and hence also issues a river. Two Philosophers stand and discourse and on the left side of the Tree.

Another Similitude.

HERMES the greatest workman and the first Master of this Art saith that the water of the Air which is between the Heavens and the Earth is the life of everything. For by the means of those two particular and natural qualities hot and moist, it unites those two contrary Elements the water and the fire as necessary mediators to agree those 2 extremes and

the heaven begins to close itself as soon upon the
earth as this water is infused from above as it were
with a fruitful Seed injected into the work of her
Womb by which means she hath conceived a sweetness
as it were of honey and a certain humidity, which
causeth it to produce diverse of colours and fruits
from whence there is yet risen reason and ground as
it were by a lineal Succession in the trace of their
Secret Ways a tree of admirable height and greatness
with a silver-like body which extends itself Largely
through the places and quarters of the world,
Whereupon the Branches of this Tree diverse sorts of
birds did repose themselves. All which flew towards
the day. Afterwards there did appear great abundance
of Flowers and infinite other rare properties were
thereto to be found, for it did also bear diverse
sorts of fruits, whereof the first were as small
grains and the other is named of the Philosophers
the Foliated Earth the third was of the most pure
Gold intermixed with many fruits necessary for
health, warming that which is cold and cooling that
which is hot and that which hath contracted an
extraordinary interperate, a kind of excessive heat
rendering the dry moist and the moist dry. Softening
that which is hard, and hardening that which is
soft; now all these conditions of contrary Essences
are the most Assured pilots of the hope of our Work,
our operations being only an imitation of diverse
natures as it is said.

The body spirit make the Spirit bodified

Kill thou the quick, quicken the which dead doth
lie.

This is the Loadstone, perfect Circle upon whose Center rests the chief mastery and the beginning of the pretended End of all our Artifice. This Maxime being true that the assurance of a good beginning is no small hope to comfortable Spirits who notwithstanding they have embargened themselves yet fearing to arise safely at the Cape of good hope when they see themselves encountered with so many contrary Winds and dangerous Rocks they are enforced many times to abandon their undertakings, to better sailors than themselves. Yet notwithstanding if we encounter some sweet and precious Halcyon in the midst of our tempest we assure ourselves to be yet at the Least in the true course of our Intentions, and by this happy Augure we begin to know the Lyon by his Paws, breathing (as one saith) under the heavy burden of our greatest travails, and bravely surmounted by the hopeful Aspect of a happy and favourable beginning.

He is half way at his Journeys end

Whose good beginning was his friend.

The black key of the reciprocal mutations of these diverse forms opens the Cabinet of natural Secrets. Whereby we taste the sweetness and maturity of this fruit of the Isle of Colchos kept by the wakening Dragon and the devouring Lyon composed to the pursuite of our Work.

Our Sacrifices end well to attain

We must be patient and refuse no pain.

Salienus speaks sufficiently of the Variety and diversity of this fruit making also ample mention of an Herb which (according with diverse others) he calleth Lunary, having a stalk far differing from others drawing his root from an earthly metal ruddy in parts; but encompassed with a black colour, or spotted easy to corrupt and to disfigure itself, and being willing to forgo his ordinary force to be more fair and more perfectly renovated that changing his might with Flowers in his due time, likewise after Seventy two hours, he encounters with the Angel of Mercury changing itself into a perfect white of a most pure Lune and again converted by a longer and more violent decoction into a Gold of such an Alloy as changes into his nature a hundred parts of Mercury. Yea a more pure gold then is produced out of any mine in the whole earth. Virgil affirms as much in the 63rd. of his AEneads speaking of a tree with golden boughs which he makes his Trojan prince to encounter in his long voyage. A tree of such excellence that whensoever any branch of it was cut off, another presently supplied the place succeeding by multiplication his predecessor, or (as the Phoenix) renewing himself out of his own Ashes.

The Fifth Figure.

The same Figure as the 4th. was.

The Third Similitude.

AVICEN treating of humidity and of all the efforts thereof saith that in the beginning there must appear blackness when heat operates upon moist bodies, which is the cause that the ancient Sages without unfolding the ambiguity of their Inigmatical figures declare that they had observed from far, a tempestuous Cloud raising itself which environed all the earth and moistened it. They say also they have forseen the great tempestuousness of the Sea and the abundant concourse of waters flowing upon the face of the earth in such sort that the form and the matter destitute of their original vigour and through complete putrefaction shall seem to threaten with obscure darkness, the King of the earth, who cries and Laments with a pitiful voice, and full of compassion. He that shall redeem me from the Servitude of this corruption shall live with me in perpetual and most happy content and reign gloriously in the sparkling clear light of my royal Throne even far excelling the precious shining of my golden Scepter.

The cloudy curtain of the mist shut up his complaint with a charming sleep; but at the break of the day there was seen to arise above the person of the King a most resplendent star and the darkness being chased away the bright shining Sun appeared in the Clouds adorned and beautified with diverse colours. The twinkling Stars did dazzle the sight and there was a perfume passing all the odours of balmes and

out of the earth proceeded a fair translucent light
with sparkling beams, In fine all which might serve
to give content or delight agreeable to the majesty
of a great King pleased with a rare noveltie.

The Sun with his golden rays and the Silver moon
encompassing this excellent beauty rendered
themselves admirable to all Spectators and this king
rapt in the contemplation of so sweet pleasures made
three fair magnificient Crowns to adorn the head of
this great beauty, the one of Iron, the other of
Silver, the third of Gold; he held in his right hand
a Sun and 7 Stars about it, which gave a great
Lustre, and in his left a golden apple and upon the
which reposed a white pidgeon which nature had also
garnished with silver and the wings of Gold.

ARISTOTLE saith that the Corruption of one thing is
the life and generation of another thing which may
be understood of the Art of our Mastery and the
preparation of corruptible humidities renovated by
this moist substance aspiring always to more and
more perfection and continuation of a much longer
life.

The Sixth Figure.

A King Crowned, with Scepter in right hand, and a
Bird standing on a Globe in his left. A King crowned
swimming in a River. The Sun appearing on the top of
a Mountain.

The Seventh Figure.

A Crowned Queen in a white imbrodered Garment, with Wings full of half Moons, and a Star on her Crown. A naked boy coming out of a brown muddy place.

The Fourth Similitude.

MENALOUS shews evidently the necessity and strict commerce the living things have with dead, in these Words: I will and look (saith he) that all those that addict themselves to our serious study and the desire to follow the same, absolute stop and order those we ourselves have kept, the better to obtain their desire, must take such a course that spiritual things may be corporated and corporeal things Spiritualized and that by a reciprocal commission and destruction of their first forms to the end to acquire a form more excellent having raised themselves from the death of putrefaction much glorious than before, by one only means and light decoction.

Many other of the best Philosophers being at Unity in this opinion do all agree in these or the like words.

Dissolve and congeal in saying.

If thou dissolve the fixt and make it fly

And fix the bird thou shalt live happily.

Or as the fountain of Learning Lovers says:

Lighten earth give weight to the fire

And thou hast that which thou canst desire.

As we have heretofore proved by diverse passages in Imitation of Senior, who invites us, as all the rest do to the necessary things of contrary natures. The Spirit saith he hath freed the body and by this deliverance the soul is extracted out of the Body and after the selfsame body is reduced into soul, the soul then changing itself into Spirit becomes a new body, for if it remains firm in the body, and that it hath renewed the body with his terrestrial and massive grossness spiritualized by the operation of this Spirit that is the perfection of our operation. Also if the same happen not to the metalline Bodys and if they loose not their first and natural being to resume a greater Luster and perfection by the Work, their first matter being destroyed by Corruption and another introduced by generation, our Labour and watching is all but vain and our oil but blown away with the Wind.

An unfortunate man fallen from the Sweet Zephirus of
his happiness and lighting by doleful disaster, into
a most filthy bay-stall, appearing as black as an
absolute moore panting and quite breathless with
much struggling, to free himself, attempting all
possible means to save his life and deliver his
fettered body from the Infected prison of this muddy
Dungeon, but his too feeble power could not second
this vote of his desire for his release out of this
place.

And seeing himself to have in Vain importuned the
heavens with Cries, to give assistance to his
Industry for the disentangling of himself out of
this poisonous Den.

In this misery he had leisure enough to attend the
last stroke of an exorable death out of all hope of
the favourable succour of any benevelent Soul full
of charity that might be drawn to a piteous
compassion of his piteous estate resolving himself
as out of dire constraint sorrowfully to finish the
abridgement of his days fatally followed at the
heels by the dismal disasters of this unclean and
uncomfortable Condition, then finding every one deaf
to his complaints shewing in his behalf hearts
harder than a senseless work.

Of a desired health the hope now being vain

His hert expecting nought but death to rid his pain.

To purpose there appears a Lady young and fair

That gave her handy help to him in this despaire.

This Lady was exceeding beautiful, both of body and face arrayed in most precious apparel of divers Colours having fair white plumes but speckled as are the peacocks which spread themselves equally upon her back by the industry of a gracious wind and favourable Zephirous the pinions of her Wings were of pure Gold interlaced with a many fair small grains of Silver upon her head (comely supported) she had a most fair Crown of Gold in the highest part whereof there shined a Star of Silver. About her neck she had a Crescent of Gold. Wherein was richly incased a precious Ruby with excellent Artistry; the true value whereof could not have been paid by the whole revenues of a most puissant King. She had also upon her feet shoes embroidered with Goldsmiths Work and from her proceeded a most fragrant smell of a most sweet and odiferous perfume.

As soon as did she perceive this poor desolate and abandoned Wretch with a Joyful countenance and amiable Aspect she offered her hand and relieved his extreme feebleness, now so destitute of his former strength that he could no more support himself or sustain his weak body, from sinking to the earth in this eminent peril of life not expecting or relying upon anything save the old proverb:

Nullam sperare Salutem.[1]

---

[1] No hope for salvation. -PNW

This Lady seeing the feebleness of this Languishing Wretch advances herself towards him and graciously drawing him unto her of the infection she Washes him neatly and presents him with a fair robe of purple and carries him with her to heaven.

SENIOR entreating of the same subject speaks alike or more clearly there is (saith he) a certain thing which is no more mortal being once revived and renewed by multiplication.

The Eighth Figure.

A King stands upon the Sun and the Sun upon his Crown, his Scepter being wrapt in a Scrowle. A Queen stands on a half Moon, and the full Moon on her head with a Scrowle in her left hand.

The Fifth Similitude.

The Philosophers to leave nothing unmentioned which they might honestly discover of this Art, do attribute unto it two bodies, viz. The Sun, and the Moon, which they call earth and water; these two bodys are also called man and woman, the which engender four Children, two little males which they call heat and cold and two little females, dryness and moisture; out of these four qualities proceedeth

a quintessence which is the white Magnesia which
doth not carry any face of falsity, and Senior
pursuing more at Large the same figure concluding in
this manner when saith he; these five are come
together and are conjoined in one, the Natural Stone
is then made of all these equal mixtures which is
called DIANA. AVICIN to the same purpose saith that
if we can arrive to this fifth thing we shall obtain
that which all the Authors call the Soul of the
World. The philosophers under the bark of this
similitude expound unto us the model and verity of
their essence by the demonstration of an Egg because
in the whole substance composed and joined together;
the first of which is the Shell, signifying the
earth, and the white is the Water. But the skin
which is between the White and the Shell is the Air
which divides the earth and the Water; the Yolk is
the fire wrapped up in a most delicate film which is
the most subtil Air, the which is the most interiour
of the most subtile, for it is of near nature and
Affinity than the fire repulsing the fire and the
Water, into the midst of the yolk, which is the
quintessential matter of the which is formed and
engendered the Chicken that afterwards doth increase
and grow, so which an Egg is all the force and
vigour together which the matter the which nature
taketh to accomplish and perfect her operation; So
is it likewise necessary that all these be found in
perfection in this our operation.

The Ninth Figure.

A body with two heads, the right yellow and left white, and both young faces, with wings, the right red, the left white, in Black clothes, and something like a platter in his right hand.

The Sixth Similitude.

The discourse of the most discreete are altogether ambiguous always intermingling their grand Writings with some kind of obscurity each of them so well understanding the other by this kind of discourse, sto that the Secret is rio more divulged by the latter then by the former.

As appears by ROSINUS conformable in the point to the old Philosophers in the Explanation of his Aenigma, concerning this matter saying that by the face which he had seen of a dead person mutilated in many pieces his body and all his limbs divided the one from the other. But the trunk of the body as yet whole, appeared as White as Salt, the head being separated from the other members, was of find Gold near unto which there was a man of very great blackness, and was composed of his limbs, of a ghastly Countenance and a hideous Aspect, who stood upright, his face turned towards the dead Corpse, in his right hand a two-edged sword imbedded in blood, in the effusion whereof (like a syrup & clear) he took his chief delight, and his most pleasant Sport, was to commit willful murder and to put all manner of persons to violent death even in cold blood. In his left hand was a scrowle containing these sayings:

I have murdered thee and cut thy body in pieces to the end so beautify thee and to make thee live a longer and happier life than thou did before did'st death conspiring against thee by the Edge of this my Sword. But I will hide thy head that thou maist not be known of men and that they may not see that in the same Equipage of mortality thou wert before, I will mingle all thy members in an earthen Vessel, where being buried and in a small time brought to Corruption that hou maist be again revived and multiplied so to produce and being forth better fruits.

The Tenth Figure.

A man with a black face, with a sword on his right hand, having cut off the legs, Arms and Head of a Man, from the trunk, the head (being of gold) he holds in his left hand. A River running by him, at the far end a ship, and several persons on the Shore looking after it.

The Seventh Similitude.

The Works of Ovid that excellent Poet and grand
Philosopher perswadeth us to esteem well of his
judgement and true knowledge and great experience he
had of the miraculous effort of our Magnesia
declaring to us the prudent wisdom of those Ancient
Sages being desirous to renovate their sliding and
decaying age, how ingeneously they did arm
themselves with a sovereign Antidote and Counter
poison at the invenomed darts of the fierce
Eumenides, the cruel plagues of life, sickness, and
old age, and being careful of the Conservation of
mankind devised a way by the voluntary dismembering
of their Bodies but cutting them in pieces, to have
them boiled to a perfect and sufficient decoction,
thereby to change the feeble Constitution of their
aged bodies to its former natural estate of youth
and vigour making themselves by dying to resurge
more strong and healthful and by disjointing their
Limbs to reunite and knit them more strong, and
firmly together.

What is that property of that nature

that doth bring to pass this operation.

The Fourth Treatise

The Prince of the Peripateticke Philosophy and great
Inquisitor of natural Secrets and curiosities saith
in his book of Generation and Corruption that man
and Seed produceth man; it being most plain that
everything Engenders his like by the animated force
and particular secret of every Seed which give every
Living form in several Essences by diverse and
sundry means but principally by the operation and
temperate heat of the Sun, without whose infused
Aids, and immediate assistance this operation could
not shew its effect. The most regular Philosophers
following the perfect pattern of wise nature, are
constrained to beg her succour and favour to their
designs in the research of their Work which cannot
be done without the mutual borrowed Aide of nature &
Art.

Perfect in all points no one thing is found

Unless it take some help from other grounds.

So saith nature to Art in her Complaints.

If thou help me I will thy work assist

If either fail the other all is mist.

For if the Artist second not the Design of nature
(although she be now so full of good intention) yet
cannot she bring to the perfect Light, nor raise
them to the highest of their absolute perfection
without their pain, patience and diligence. Neither

can all our Artifice at all perfect in those vain Searches, but remain unfruitful and unfittable without we have the favourable Assistance of Nature.

This plainly appears unto us that they are always to mutually aid one another and that our Art ought to govern the heat by the temperature of the Sun -to produce this Stone. But the Event and good success of all these things ought to be extracted by our Sage Imitators in seven several manners which doth open us the door graciously to induct us to the full understanding of the perfect heats.

The Eleventh Figure.

A man stands in a Cauldren with a Bird upon his head. A young man bloweth the fire under it, at the bottom, a passage for the liquor to run into a Glass.

First we must of necessity practise such a kind of heat as may mollifie and melt the hardness of the earth, seething together the gross and the hard by a temperate fire of Corruption which is the beginning of all the Work, confirmed by the best Authors.

If it be not putrefied it can neither dissolve nor melt and if it be not dissolved it comes to nothing, saith MORIEN and PLATO. Note that without corruption there can be no putrefaction. Therefore saith he to attain to putrefaction must be the principal Aim of all our intentions. Accordingly the same philosopher declares that he had never seen any living Creature bred without putrefaction and Urging it farther in vain were the Work of Alchemy without you did first putrefy.

PARAMENIDES also exclaims the same in these Words: If the body be not destroyed, demolished, all rotted and totally corrupted by putrefaction the hidden and secret virtue of the matter could never be drawn out, nor perfectly conjoined to the Body. The great Rosary maintains the same opinion to be most assured and infallible as appears in this metaphorical figure. We hold saith he for a most true MAXIME that the head of our Art is a Crow flying, without Wings in the obscurity of the night as well as in the light of the day. But by some means this must be done, SOCRATES gives us very good advice speaking - thus of the first heat agreeable to corruption. The Vents and little holes which are the breathing places, and pores of the earth shall open and dissolves to the end she may retain into herself the force and vigour as well of the fire as of the waters.

The Twelfth Figure.

Three Birds in a Vessel, one black falling on his back, another white, and another red, both pecking at the black one, and he at them. The Vessel standing as it were upon beams, round at the top, and flames persuing throughout.

Secondly by such a heat is necessary for us by the virtue whereof the darkness may be expulsed from the earth according to the proverb of SENIOR, heat saith he makes all things White and all White things do afterwards become red.

The Water likewise by his Virtue brings Whiteness which the fire shortly after illuminates. But the Center doth then pierce and shines through the subtilized earth like a Ruby by the tinging Spirit of the fire, to the which also agrees the Authority of SOCRATES: Rejoyce thy heart when thou seest an admirable light Issue from dark obscurity.

The Thirteenth Figure.

A Griffon in a Vessel; and a naked boy blowing into his mouth with a pair of Bellows. Round the bottom of the Vessel Beams. The Vessel Crowned and flames issuing out of it.

Thirdly heat rightly disposed directs everything in the height of his perfection by the Secret force, with the which she animates Bodies by the means of their Corruption. For which Cause MORIEN saith that nothing animates itself, but after putrefaction and that all the force of the Mastery avails not if this Corruption do not preceede as it is assuredly affirmed to us by the Turba Philosophorum[2]; who by a

common consent attribute to this heat the
Jurisdiction and power to animate bodies in giving
them a living essence after this Putrefaction to
fill with waterish humours that which was formerly
firm and solid, or other Like, or contrary
operations because heat continues this properly both
to fix and to resolve and here lyeth the knot of the
business. In which absolutely consists the
perfection of the Workman. Therefore ought we to lay
fast hold in this assured precept that rather to get
the comfortable hope of being able to attain the
precious prize and expected Salary of our black
earth, the dissolved, and congealed, so often
remembered by the best Authors and etc. and so often
repeated by us. It is no small matter to know the
fire that causeth this putrefaction and many other
diverse and fair Effects, upon which the entrance
and conclusion of our Saturne wholly depends.

If thou this work will speedily conclude

Make the fixt fly, soften the hard and rude.

Because the essence of this Mastery takes his force
from contrary qualities perfectly united. RASIS in
his Treatise of Lights affirms as much speaking of
the necessity of this metalline mixture. No man
saith he can reduce a heavy thing to Lightness
without the help of a thing that is light, no more
than he can transmute a light thing into a heavy,
without the Intermission of a heavy Body.

The Fourteeneth Figure.

A Griffon with 2 heads enclosed in the Vessel. The Vessel Crowned at the top and stopped.

Fourthly heat purifies, driving from his heat the least object of Impurity; CALID says we must waste the matter with a hot fire, if we will make an apparent mutation we must also know that the minerals joined and knit together easily depart from their first habitation by the recipical communication of their proper influence by the infusion equally disposed through the total mass of their community despoiling themselves of a particular Vestment, to make a White; after an equal proportion and measure to all the mineral substance so quitting their evil infected savour by means of our renewed Elixir.

Of which HERMES treats very well to the purpose saying that it is most necessary to separate the Gross from the Subtill, the earth from the fire, the thick from the thin. And it is very moot for me to report in this place the conceit of ALPHYDIUS in his Treatise which contradicts not in any thing that we have spoken. You shall know by exact reading of his learned writings the same advice given by so many renowned Authors which brought us out of doubt in this way: The earth saith he melteth itself as a water out of which proceeds a fire, yea for the earth contains in itself a fire, as the Air is contained in the Water. RASIS also adviseth us that a certain Artifice ought to preceed the perfect operation, which we fitly call mundification because we must resolve, to make the matter more tractable and that it be reduced to Water which is supple and principle of all things, for of water all things are made, the which is done only by Putrefaction, for which the beginning of this mundification and may get an assured prognostick of the success of this Stone of the Sages; If the filthy and deformed parts, like Excrements hurtful and superfluous to the purity of this fair work, be entirely separated and excluded.

The Fifteenth Figure.

An Eagle with 3 heads Crowned in a Vessel stands upon leaves (beams?). The top of the Vessel crowned out of which issues Flames.

Filthy heat is elevated by the Virtue of the fire and the hidden spirit of the earth is returned into Air as HERMES saith in the Smaragdine Table, in these words, it gently ascends, from earth to Heaven, and again descends from Heaven to earth, so obtaining the virtue of both and a power above every power and in another place subtilize the gross, and make the subtil gross, and thou shalt have the glory of the World. RIPLEY in his 12 Gates says the same in another figure.

Gate Three.

Raise saith he the birds out of their nest

And then again return them to their rest.

Which is only to draw the Spirit from the earth, and return it thither again, and to this purpose say the Philosophers that they acknowledge him for a Master of this Science that knows how to draw light from out of obscurity. MORIEN falling upon the same point (to whose sweet Concordance we strain and tune all our harmony) composed from the brains of so many Sundry and transcending agreeing Spirits concluding it for an absolute truth: He that can give Solace to the Soul by drawing it out of Putrefaction knows the greatest Secret of this Mystery, herein ALPHIDIUS agrees, cause saith he this vapour to ascend else you effect nothing.

The Sixteenth Figure.

A Peacock with a spread Tail in a Vessel, crowned at the Top and Stopped.

Sixthly that when the heat is so much multiplied in the earth that the strongest parts be reduced and united together, and rendered more volatile, surpassing Impurity and all the other Elements. But this heat must only be augmented, to the equality and proportion of the coldness of the Swan (?). CALIDS words do proud what we say, extinguish saith he the fire of one thing with the coldness of the other things, yet advisedly, let this coldness exceed the natural heat, but one only degree for

fear of an entire Suffocation as Raymond testifies
in the Theory of his Testament.

The Seventeenth Figure.

A Queen Crowned, (with a Globe in her right hand,
and Scepter in her left, standing upon a half moon,
with a Ray round about her body) in the Vessel, the
top crowned and flames issuing out thereof.

And seventhly heat mortifies the cold earth to the
which the saying of SOCRATES doth well agree, in
these words; then when the heat penetrates it doth
cause those things that are gross, and terrestrial
to become subtile, and spiritual accommodating
themselves to the matter rather than to the final
form, continually working by means of the aforesaid
heat. And this is the which the Philosophers mean
when candidly they speak of distilling 7 times,
understanding by 7 Colours, applying by the
continual decoction in one only Vessel without once
touching the same lending all to nature which
separates and mingles them together according to her
own balance and not ours.

The Eighteenth Figure.

A King Crowned in the Vessel stands upon a half moon
reverse, with a Scepter in his right hand, and a
Globe in his left, the top crowned and stopped.

Actor in the fourth of his Problems gives us another Instruction fittingly to temper and govern the heat of the fire necessary to the operation in these terms, than which the Sun is restrained which is as much as debilitated and returned into his first matter he shews the first degree which is unto us a true sign of infirmity, principally because of the diminution of his natural heat, being then in his blackness, then is a way by the breath of the Lyon, to corrupt this first natural heat, augmenting it to a burning fire more digesting then common fire and this excessive Ardour demonstrates the second degree, which proceeds from the admirable great heat of the fire, whereby we understand Putrefaction which is the privation of the Form. And again a Plain other positure of the Air of the 3rd. degree follows at the heels of the two others not burning but of temperate quality with a mediocrity of Air, and order more regular and changes all Violence into a tranquility; here you may behold the true means to give an End to the Work and an assured beaten path, to the culture of the hopeful Vine and the achieving which good success the Comfort of a Delicious Air of health and prosperity.

The Fifth Treatise.

The Whole Operation comprized in four brief Articles, easy to be understood.

The First Article.

The first step established by the true Alchymists to mount the golden Scale of our happy work is called by the most expert in this Hermetical Art Solution, which requires (according to nature itself) that the body should be boiled till it comes to perfect decoction. All our Mastery being no other thing than boiling.

Seeth, Seeth, and again Seeth and let it not seem tedious unto thee for the more thou seethest the more thou dissolvest, the more thou seethest the more thou whitenest, the more thou seethest, the more thou reddest. In fine decoct at the beginning, decoct at the middle, and decoct in the End; Seeing this Art consists in nothing but decoction. But the decoction of thy material must be perfected in one only water that is in our Quicksilver which serves us as the matter and in one only Sulphur which is the form.

Hereby we must clearly understand that the vital Silver which cleaves itself, doth firmly adhere and is annexed to the dissolving Sulphur. Join the dry with the moist and then hath the Mastery; convert water into fire and the dry into moist. In brief the Elements one into another, and you have a firm passage unto all that your heart can desire in this Art. *Converte Elementa and quod quaris invenies.*[3]

The best understandeth perfectly all happiness to be at your disposing if you know the means to join Mercury and Sulphur together. Now this Solution is no other thing in a plain order inconjoining humidity with dryness properly called Putrefaction, which it totally corrupts the matter and brings it to blackness. MORIEN attributes to it the same effect upon the same near Putrefaction, wherein we comfort our hopes in the Work, this being the key that opens all the Locks of the very Heaven of the Philosophers; If it be not (saith he) putrified and black it will not dissolve and if it dissolve not the water cannot pierce through all the body as it ought necessarily, nor penetrate nor blanch. It must therefore dye to revive as the grain of Corn which neither germinates nor bringeth forth profit, if first it do not well dye altogether.

The Nineteeneth Figure.

Two women beating and raceing Clopex near a fountain.

---

[3] Turn the elements and find out why. -PNW

The Second Article.

The second step is called Coagulation, which notwithstanding may be said to be one and the same thing with Solution working the same effects. The difference that is between them is only caused by a small and almost insensible distance in the perfecting the mutation of the first Essences into diverse natures qualified, which diverse names only to oppose the confusion of the first Intentions, and utterly to deprive the Ignorant from apprehending the Secret, and to lead the Children of Art by the hand to the true understanding of the same.

This Coagulation then doth again renovate the water in the Body, for in congealing it dissolves, and in dissolving it congeals, to shew us that the Quick Silver which is the dissolution of the Metallick Sulphur and (which he draws unto himself thereby to be congealed) desires again to be rejoined to the radical humidity of this Sulphur and this Sulphur as greedily seeks to be again conjoined to his Mercury by which reciprocal amity it is easily perceived that the one cannot live without the other, embracing amiably the other as being indeed one only nature as most learnedly is published by CALID under the name of all the Philosophers, in his book of the Secrets of Alchemy, saying:

Nature rejoiceth in Nature, nature overcomes nature, nature retaineth nature, nature whiteneth nature, and nature rubifies and afterwards he addeth; Generation is fixed with Generation, and Generation

is virtuous with Generation. By good right then say
we that our Mercury seeks always the fellowship of
the Sulphur to serve him as his formed from whom he
had before him separated with so many unutterable
signs and tears as not being able to suffer the
dissolution of two so perfect Lovers, for this
Sulphur which is the form of Mercury makes him
return again to him, drawing him from the water of
the earth as soon as they are disunited, to the end
that of this body composed of matter which is
Mercury and of form which is Sulphur, we may extract
a perfect essence in the which are to be seen a
happy diversity of Colours because the property of
the Working things no sooner begins to alter, but
the pure conduit and exquisite achievement of living
and animated things wisely governed and learnedly
disposed, by the head and hands of the expert who
have already guided the Elements.

It being no small matter to find a good Pilot that
can securely travel in these Seas although he be
provided of an able Vessel that is to say working
upon the true matter, and knowledge, or else the
quantities and qualities of the operation of things,
because that in Solution the Mercury is become the
Agent, whereas in Coagulation, it is the patient of
the operation which happens. And here we may
ourselves apprehend that this Science if fitly
compared to the Sport of little Children for only
Art is called play, but principally that of Letters
which is called LUDAS LITERARUM. In which the best
Spirits take greatest pleasure and the Learned as
much content as Children delights in frivolous
pastimes, not apprehending the least Inconvenience,
as this present figure represents unto us.

The Twentieth Figure

The boys play, and riding on a Hobby-horse, another scourging a Top, and a girl dancing and leaping between them.

The Third Article.

The third degree of the Naturalists is Sublimation, by which the gross and massie earth changes itself into his contrary which is humidity and then may it easily be distilled, after it hath encompassed this Condition for as soon as the water is reduced and brought to influxtion into his proper earth it begins then to retain the quality of the Air lifting up itself by little and Little, and puffing up the earth, kept till then in quiet repose by the thirsty siccity like a compact body and much pressed together the which nevertheless resumes there his Spirits and extends himself more Largely by the Influence of the humour wherewith he is imbibed and entertains himself by the Infusion in this Solid Body in form of a porous Cloud, and like to the Water that swims in the upper part of an Egg that is to say the Soul of the Quintessence which we naturally call Tincture, Ferment, Soul, Oil; because it is the matter most necessary and in the nearest degree to the Stone of the Wise men, for as much as by this Sublimation, cinders are produced the which perfectly (but above all by the Assistance of God, without whose favour nothing proves fortunate) arrogates to himself the limits and measures of the fire in which it is enclosed as it were with natural compeers firmly shut up: Ripley in the same sense agrees with us, saying, make fire in thy glass, within (viz.) in the earth, where the natural fire is enclosed.

This brief method whereof we have given you liberal instruction, seems unto me to be the shortest way and the true philosophical sublimation to reach to the perfection of this grand and rare labour, aptly to the purpose compared for the purity and admirable Candor thereof, to the ordinary business of women which is their Laundry which hath this property to make things infinitely white, which before appeared to be sluttish and foul as by this figure you may perfectly perceive. But I must also advise you that I am not alone in this Allusion, there being nothing so common in the best Authors, as to call it womans work, and childrens play. Children using continually to be mire and be soiled in the ordure of their Excrements which represents unto us this blackness drawn out of the proper natural mixtures of our mineral body without any other operation or Artistry then of his fire, hot and moist digestion, and vaporous, the which blackness and putrefaction, is afterwards cleansed by the Whiteness that comes to take place making the house neat and clean from all former Imperfections with the same lye and clear Water the woman takes to wash the Child and purify him for his more entire preservation.

The Twenty-first Figure.

A Sun rising behind the Top of a Mountain, A man going towards it, and by him runs a River.

The Fourth Article.

This last of our Articles advertizeth the reader that the water should be separated and divided from the Earth and afterwards rejoined again to the end that these two bodies being straightly united may again be one homogeneal thing, and so firmly and fast knit together, that no more they may be any separation. Such must also be the intention of the workman, otherwise his Labour vainly undertaken will now come to an end but remaining always in an estate in like to leave nothing to the Artist but a carefull and mournful remorse to find themselves the Servants of Ignorance in not being able to reduce this work into the natural union of a body composed of diverse things, and differing in quality of the which necessarily he must serve himself to the raising of this rare edifice.

Neither more or less than the wise Architect that frames a building of diverse materials, whose Idea notwithstanding the Varieties aims at one only end, a palace or a house and structure of diverse parts steadfastly united in one body composed of diverse Instruments.

Then the which may be said of our Composition and of the proportion thereon to be observed, is succinctly comprised in the brief method of these four precedent Articles without otherwise Intricating the Spirits almost already amazed and confounded with the Intricate paths, doubtful stops and hyperbolical discourses of so many Authors, which speak of it but mystically in such art as they draw into Error, such as are less advised, under the doubtful veil of so many obscurities, and cause them to run headlong into the Unbottomed Pit, as soon as the Shining Sun illuminates with his rays some part of the Superficies, so that having already promised to themselves so many Golden mountains upon this smiling fortune following then all panting and out of breath, their business thinking to surprise and snatch the Moon in their teeth, whereof they are driven to repent of the sun light of their inconsiderate rashness;

Odj pupillos praecocis Ingeng.

No man doth err so much in heat or cold as doth the Artist, that is rash or bold.

Patience is a fair guide to bring a man to the end of the most hard Attempts, for most in request are the things most difficult. This is the cause that TURBA bids us so often be diligent and patient, and not to be offended, with the tediousness.

And AUGURELL:

Patience will be thy true and faithful guide
Consule with her and thou shalt never slide.

The Twenty-second Figure.

A full Moon rising at the top of a mountain, the
lower part of her face ruddy, the top part white.

Of the Order and Regiment of the Fire.

After all these Articles we are to treat of the true
manner, well and methodically to govern the fire in
the due proportion of his degrees, the knowledge
whereof is to us so necessary, that without this
Science all our operations shall prove but
unprofitable.

For although we are assured of the right choice of
the matter and so also understand the means to sow
it in the proper earth, yet all is nothing.

He that wants one thing wanteth all the rest
Our fire unknown, our hope is quite disperst.

As the least Vice, or defect breeds more disgrace to the most generous Spirit then all his virtues can give him praise and reputation and therefore 'tis premptorily spoken of.

The wise Inquisitor must doubt of nought

All that he wants stands ready in his thought

One regimen of fire makes perfect all,

Guiding your feet so you cannot fall.

That is the faithful Agent must dispose

The work from the beginning to the Close

He is the Guardian of our Cittadell

And saves his King by standing Sentenal.

PONTANUS gives us good instruction to this purpose, in an Epistle of his, making us happy by his mishaps (if other mens faults may be our warnings) who by his own defect was carried quite out of the light of his designs being not able to advance his Work beyond the beginning in two hundred times that he had attempted it, notwithstanding that his building was raised upon a due and true foundation. This Ignorance cost him deere both in time and expence, and bred much repentance though he were guarded with all the patience that is required. But the natural fire necessary to this fair business, giving him no Assistance he was disappointed of his expected desires so often as he persisted in his former Course, much power hath MERCURY the father of the family instructing and governing this rich Vessel much might here be discussed but our pen hath not permission to write plainer, when a thing is addressed to heat it ought to be in such a manner as there may not be any perceivable motion at all, but only an insensible change of his natural order; agreeable to the Sun, whose heat we ought especially to Imitate, which is as much as if we should say unto you that a terrestrial thing without Spirit, may be animated by the means of a natural heat, conformable to that of the Sun and Moon, not excessive or scorching but only moderate and according with a well-tempered body. Now of what kind of qualities these two principal Celestial Luminaries be; SENIOR shews unto us when he says, that the Sun is of a moderate heat, and the Moon cold and moist, but as less perfect she mounts up desirous of better state, and borrowing from the more noble party, that which she wanteth, until in the end she appeareth of as much Vigour and virtue as he had that favourably communicated to her, So that soon after they both put equal Agents, upon Bodys with their Celestial influence and do abundantly replenish them with their fain Illuminated motion as heat and moisture, cause generation they are therefore most necessary to our design, as all Authors affirm upon which FLAMEL grounding himself in his Philosophical

Summary:

"That heat and this humidity

is nourishment in Verity,

To all things this world brings forth

Having Life much or little worth,

As Minerals and Vegetables

Yea animals and rationals;

This heat is no burnt Coal or Wood

They do much hurt, but little good.

They are too full of violence.

Not nourishing but breed offence

But it must be a warming heat

Wherein kind moisture hath his seat,

Like to the Sun that comforts all

Else will your comfort be but small."

Thus we have sufficiently declared the Mastery of the Ancients how by the renovation of these two means, we hope to obtain the glittering beams of the radiant Sun coming to refresh his amourous ardour in the Silver Bosome of the depurated Moon, from whence we shall see the issue, a thousand little Suns, that is to say, infinite and which may be multiplied without end or number. This being now the Stone of the Sages. Scala Philosophorum to advance this excellent knowledge entirely describes what should be the fire of our Mastery and with what temperature the Soul of the Philosophers would be entertained.

We will produce as by the way some diversities of opinions, It is well said in the place above mentioned that the heat or fire requisite to this work is comprised in one only form but it is too successively spoken, because:

When I would use in Art a brievity

The sense is lost in my obscurity.

We will therefore clear ourselves of this doubt in speaking plainly that some of the Turba will that the heat of the first address, or regiment, should in some sort have relation to the heat of a hatching Hen, others will have it resemble the heat of a human body, even such as the perfect coction or digestion of meat in the Stomach changing into the substance of the body the necessary quantity of the nourishing thing. Others will have it equal to the heat of the Sun which (according to the object wherewith it encounters) produces contrary effects although inalterable in his own nature, as doth our Stone aforesaid which without any Labour is brought to perfection, changing his first being and suffering himself to die that he may again revive by the aid of that which caused his death.

Because the fire of the Philosophers retains the effects of the Scorpion, which carries in himself

Life and Death, killing by his Poison and being
applied to the Wound becomes a sovereign salve. The
too violent fire ruins that which it encounters, the
moderate refreshes and insensibly comforts that
which he would help and relieve with his humidity as
Calid says the Lesser fire grinds all things, and
this is the hopeful means of a praiseworthy end,
from the beginning of the interprised Work, to
minister a temperate heat that which without burning
penetrates vigoursly into the entrails of his
massive body that she softens his hardness, and
makes him comply to all his pleasure as the Water
which by long continuance of his dripping wears out
and pierces the most Solid Rocks, which by an open
form he were now able to effect the matter altered,
and gently chased retains no more his Luster, but
patentially and changing his fair tincture covers
himself with an obscure Vail infinitely black which
makes him as it were Lepreous and Corrupted in all
the parts of his body. As the Fountain of Lovers
calls it mesea Gold, or the lead of the
philosophers.

His former state he seeks to change

His Coal black his visage strange.

But the all producing times dissipates in the Second
change, the shady darkness and in due Season
withdraws the body from the black dens of his long
endured prison, redelivering unto him a new form
freed from this Corruption, from which being
cleansed he resumes the agreeable face of his
perfection.

Now this black Sun burnt Indian turned into a most white Swan.

The true heat requisite for this purpose should be neither more or less burning than of the Sun, that is moderate and temperate, because the gentle fire gives hope of health and perfects all things as the Turba affirms.

But the heat necessary in the alterative principles of our operation is in the Sign of Gemini and when the Colours are become white the multiplication doth appear with an absolute dryness in the Stone. Now to know whether this especial Sign does rule or not, we have no way so good to decide our Judgement as to examine whether our heat be the same that is in the Sun; For it is that only we desire, for the great Sympathy is in them both, and disagree most in the same, changing themselves according to the Signs which are predominant, more violent or more gentle, naturally notwithstanding and without any artifice.

But as soon as the Stone is dryed and may be reduced into powder, the fire hitherto having been moderate, ought to reinforce himself and to act upon this body more forcibly. To the end that by his augmented Ardour, he may be made to change his habit, to put off his white robe and to put on a robe of a higher colour, more transparent and Vermillion like which is the ordinary Rubys and right rich Vesture of our great King, now delivered from the prison (wherein so long a season he hath lived in so great and grievous an endurance) by the great diligence of his faithful governour who hath cherished him. The degree of his heat is the same that swayeth in the Sign of the Lyon more furious and flaming than all the other Signs of the Zodiac, for then is the Sun most vehement as in his highest degree of heat and elevated into the supreme dignity of his Celestial domicile.

This is sufficiently handled compendiously (which we effect) in this our Philosophical Instruction, the way to be kept, and strictly observed in the government of the fire of the Philosophers without the which travelist in vain wheresoever thou beest, thou wouldest make an Assay of this last piece, wherein consists the whole perfection of this absolute Work; here we have Laid all before thee more clearly than if our discourse were delivered with a larger plixity of Speech. If thou understandest me I have said enough. By the paws you may know the Lyon and the workman by his description of the Work.

Of the Colours successively appearing in the preparation of the Stone.

Many Authors writing of this Herculean Labour may seem to contradict and overthrow one another in the diversities of their opinions, and if that we do not more expertly examine their common Intentions or if we are not well advised of their purpose in this ambiguity, we may sweat a long time in the

extraction of the Spirit of these their curious
Subtleties so intricate are their ambiguous
Writings, that it is infinitely difficult to Atomize
into all their parts and chiefly when they treat of
the colours in the Work of the which we will
Succinctly say something; yet I will not adventure
to bring them all to light and fetch them one
another after another out of their Den's believing
ourself sufficiently discharged of our undertaken
promise, if we produce the most apparent and those
which contain the others (the rest being guided with
too slight a Consequence) as to manifest the Secret
of the principal points, and which have managed the
whole Economy and the most weighty business of their
Lord by whose Intelligence we shall have assured
knowledge of all that is hid even in the most Secret
and Sacred Cabinet of this King so expedient
sagefull in this business that without Inquiring
after the Inferiour offices of the Cabinet? Of the
dignities and qualities the Officers may attain by
means of those Colours. Miraldus one of the Turba
saith to this purpose (consenting with all other
good authors) that our metalline Body becomes twice
black, twice White, and also twice red, which be the
principal permanent Colours: changing by the more or
less measure of heat, for it is most plain that
there is an Infinite number of others.

But because they are only Accidental we do not rank
them in the list of our Accompt, for fear of
confounding light brains, as well as our Writings
and that as many Colours as possibly can be Imagined
do wholly depend upon these three above mentioned,
and return in the End by a proportioned Symmetry to
one or other of the designated and , it is not

without reason that the Authors by the Inspiration of some holy rapture or revelation do abridge this diversely, to the Divine mystical Ternary number which meets, as in a Center point the glorious termination of all Felicity. Amongst these three notwithstanding (to conceal nothing of our brief method) which are the principal and permanent of our terrestrial and metallick king of Philosophers, we may also well discern some other different and intermixed the which notwithstanding purposely and out of good reason we conclude as being but imperfect Colours and not of such nature and consistent as they should be worthy to be reckoned amongst those three that are more permanent, the black, the white, and the red. The which Immediately and absolutely comprehend all the Accidental; Therefore is it needless to write of them any further unless to content the Curious; we have now already produced the causes to the means as honestly to pass on in Silence the general number of those which successively appear the one after the other between the principal above named because the effects are of small success in regard of the least of the other (our natural Works acting nothing in Vain) and their colours of so small appearance gliding as it were insensibly out of sight we leave them more suddenly then they them quit us, for they pass by with so swift a march, that scarce the shadows of their substance appears when they vanish in the Vessel with a pace equal to inconstance.

This is the cause wherefore we have not discoursed of each particular species and their property. These being something else to do then to take an uncertain thing for a certain. For all those Colours come with so feeble and slow a pace that they cannot be discerned, we will not write, attending more profitable designes and speak only of the yellow Colours which come next the perfect Whiteness before the last redness because it remains a long time visible in the matter in comparison of the speed of the others and for this reason the Philosophers give him a place of principality as to the others, reckoning it in the rank of other necessary Colours, not that it yet stays so long in the vessel as the three which remain permanent in the matter the space of fourty days apiece, but for this that after those others she makes the longest abode, with 4 Colours, are compared to the four Elements which have influence and Dominion over all bodies as well humans as Animal and mineral. The black to the earth which is the Lead of the Philosophers and the firm base to support the others; The white to the water, which serves as Sperm to the Celestial woman for Generation. The yellow to the Air which is the father of life, and the red to the fire which is the end of the work and his last perfection. The black which appears twice as well as the red is in great Credit amongst the famous, because he hath the keys to open the Door of which of the Colours he pleases, having a fire that administers to him all things, yet are needful and upon which only he relieves holding the others under his Laws, for without that there is no happy effect to be expected, of all the enterprise, this humour is not so intractable, nor hard to manage as the rest, but much more handy and easy to govern and demand no other sustenance but a gentle heat which will prepare the corrupted LATIN to good obedience through patience and humility sooner than by the vigour or violence of a rash Governour which instead of mending would marr all.

Senior gives a Law in this point to many good Authors which all approve his opinion in his Writing agreeing to it, Advise that the perfect decoction of the matter should be entertained with a temperate heat until the putrefied Crow be fled, and hath yielded his place to another tincture, and seeing how it is the fire as is reported in the Complaint of Nature speaking thus:

Fire is Master of every Thing.

And causeth all things fresh to Spring;

And hath Life by heat inspired.

Which guides the work and disposes all at his pleasure as a faithfull Interpreter of the dark Language which doth direct the Work the most assured way.

I shall no more be daunted saith the Dectumn of the Turba; and have announced by the mouth of Lucas one of their associates that they held in great Estimation the workman that understands the fire and seasonably to increase it.

Take heed saith he of a fire which is too strong when you begin; For if it be too violent before the time and exceed his dimensions he will burn that which he should putrefy which is the principal of

Life. So our unprofitable Labour would yield us
nothing but Repentance, Confusion and unspeakable
displeasure, vainly expecting good by Violence
caused by Rebellion and obstinacy. To which purpose
Mary the Prophetess tells us; that the strong fire
hinders Conjunction and the true dissolution of
nature; and elsewhere she saith; the strong maketh
the White red before his time. And Trevisan says,
that the gentle and temperate fire perfects the Work
when as the violent doth utterly destroy it. If in
everything the end of the Enterprise is to be
considered at the beginning. Then in this
principally we ought to be most vigilant because if
we know not the Register of our fire in every
Season, which is the greatest happiness to our
Attempts and the only way of bringing our Work to
his perfection, our labour is lost, for in the
knowledge of the orderly progression of Colours
consists the main point of this Mystical Science and
of the tree of HERMES so often and so Divinely
celebrated in the Songs of all the Philosophers.

Know but our Brass, which if thou hittest right

Thou knowest all, whereof our pens do write

Whose power first makes black the enclosed matter

Then brings it into Water, moist as a Water

And lastly to a powder perfect red

Setting a Diademe upon thy head.

Baldus in the Turba speaking of these Colours whose
Apposition we ought strictly to observe, gives us

Advertisement to decoct our Composition until we see
it become White, the which afterwards we must quench
in Vinegar, by which means the Mineral Water, of the
matter which is the fire and water philosophical,
for our Water is fire burning the Sun, more than
fire agreeable to the Rosary and the Turba which say
that our Water is stronger than fire because it
makes the body of Gold a meer Spirit, which the fire
can never do, and with Geber; we must Learn (saith
he) to the black from the White for the white is a
Sign approaching neer to fixation.

Now we cannot better distinguish them then by a fire
of Calcination, seeing that without Addition or
Multiplication of the heat, by the gentle
temperature of the which hath proceeded and procured
this corrupt blackness of division of the degrees of
our colours, cannot easily be performed, though in
fine it may be obtained by the industry of such a
fire, and then their remains to us a gross Earth
(which many have called the father of the matter) in
the form of an Earth black and rude which is their
Saturn, a Leprous and black earth which others name
the inferiour World, the which can no more mix
itself with the pure and subtil matter of the Stone;
for we are enjoined to separate the subtil from the
gross and from the pure the impure which is by
decoction without touch of hand or foot because the
great Work dissolves itself and separates itself as
is affirmed by RAYMOND, TREVISAN, & HORTULANUS.

Upon the Smaragdine Table saith the same, you must
separate it, that is to say dissolve, for
dissolution is the separation of parts; And

whosoever knows the Art of dissolving is arrived to
the Secret according to RASIS.

Now this is the Rendezvous to which we are summoned
by all the best Philosophers, when they so often
advertise us that with the true white and the red
are to be extracted from the black and then there is
nothing to be found in him that is superfluous,
having resigned all his power to the two foresaid
Colours and he is now no longer subject to
Alteration, but yields himself afterwards
conformable, in all things to the compleate red. And
this is the Cause why they would draw him by the
vehement and force of the fire.

In the Turba it is said that when the Colours begin
more and more to enter into motion and Alteration
the fire ought to be more augmented, and be more
Vehement than before, so that hereafter we shall not
fear any danger, for the matter is fixed in the
White, at which time the Soul inseparable joins
itself to the body and the Spirit now descended from
heaven into this earth, do never depart from thence
again, which is confirmed by Lucas, when our
Magnesia (saith he) is transformed White, she
recalls the Spirit into her which had lost her, and
thence forth they now separate themselves anymore.

The Father of the Philosophers Hermes passes you
further and says, that it is not necessary to finish
the white Magnesia, until all the Colours be
accomplished that which sub-divide themselves into
four diverse Waters, that is to say, from one into

two, and three into one, the last of which parts agreeth with the heat and the 3 others with the moisture.

Hold this for assured that these foresaid Waters are the Philosophers weights, and these Weights are the colours of the matter and the 3 principal Colours are the Philosophers 3 fires, natural, not natural, and against Nature. The Comparisons that the Lovers of this Science make when they allude our Work to WINE, is not from the purpose which I might succinctly propound the less to trouble the benevolent Reader, we must understand that the Savour of the Wine, within the Juice, like as the white colour of the matter shall be drawn out of this quintessence, but his nature shall be finished in the third degree according to true proportion, for it Augments itself in the decoction and forms itself in the pulverization (?) which are the sole means to conceal the beginning and of this natural seed. For the same cause diverse Authors write that their philosophical brass shall be absolutely perfect in seven days, by which we understand the seven metallical Colours whereof the perfect red is the last. Others prolong not this time of perfection further then to four days having relation to the four principal Colours which diversities do only admit, and of which especially depends the whole Work; Others allow but 3 days which are attributed to the strongest, and most necessary Colours of the matter and some others less sparing of time and delivering it by larger measure do charitably afford us a whole year to bring the business out of towtellage and to give it absolute power and after to manage his own rights, without other Governor

then self-discretion, now capable to entertain a
world with his bounteous Liberality. And this yearly
term may well be accommodated to the four Seasons
and as it is by some of the four Elements, which
have no small right in this great matter conformable
to the judgment of Alphidius seconded with many more
of the Society, determining the End of the work by
the End of the four Quarters of the year: Spring,
Summer, Autumn & Winter, because the year is
composed of these four Seasons, many others abridge
it to a day which is the time of perfect decoction
speaking metaphorically, for a year philosophical is
the time of decoction, which some will have to be a
Week, others a month. ARNOLD, RAYMOND, GEBER,
HORTULAN & AUGURELL, testify as with 3 years
expressing for a Colour a year. All which
diversities tend but to one and the same sence, by
the doctrine expressed, and precepts of the most
ingenious understands, which are reserved in their
most secret Cabinet the exposition of the times,
weights and matter, that the Ignorant might not
understand therein by which means the Sages do
discreetly cast a Cloud before the venerable Entry
of their mysterious school, least fools should find
it, As PLATO absolutely forbid the publishing of his
Divine Doctrine, to those that had not the knowledge
of Mathematicks. It is the general Charge of all the
philosophers upon penalty that they should not
deliver their mysteries, but masked with
AEnigmatical and ambiguous Speeches, to the End
their work should only be communicated to the
Capacity of the Children of this Science and to the
diligent search of transcendent Spirits, of which
number, if they be not, they ought not to
intermeddle, but to withdraw themselves, and not to
effect, or try to set foot on the threshold of this

so perilious a Port for them, least they should make
a sotish stumble and measure the floor with their
nose.

You foolish and propharie fly far from hence

This our art loves wisdom and diligence.

The Sixth Treatise.

The disposition of the whole work and the
preparation of the Stone.

Calcination and Dealbation amongst the Philosophers,
hold the place of a good father of a family in
providing all things fitting the necessity of his
household, so do the so hold the prime degree in
this Aeconomy from the beginning of the work and the
principal Charge of the entire Administration of
metals, till that by his provident discretion the
vice Governour changing them every one into his due
place have reduced all to the honourableness of
their perfection. Now being to treat of this
Dealbation it is remarkable that the philosophers do
establish in it 3 diverse fashions, whereof the two
first appertain to the body and the third to the
Spirit. The first is the preparation of the cold
humidity, which preserves combustible matter from
the injury of the fire which they call their Saturn,
because Saturn is said to make Congealation of the
Sperms and by that preparation duly made in the
Souls we perceive the good success of a plausible
beginning. The Second is an unctuous humidity,
(which makes the combustible parts apt to retain the
fire) which is otherwise called the viscous oil,
appearing after the Corruption, this oil is that
which gives tincture and is the first philosophical
Menstrua and their first vessel. But the third is an
Jurisdiction of the dry earth that is into White
indued with a true pure fixed and subtil humidity
which yields no flame forming notwithstanding

himself into a body clear, transparent, shining and diaphanous like a glass which is pure and perfect Whiteness of the Pearl of the Philosophers and their white gold, and this is half of their Work, their Calcination being no other than pure blanching. As Morien says, when our Gold shall be Whitened after his blackness it is named our Gold, our Calx, our Magnesia and our permanent Water. See then the manner of Calcining Philosophically, which is by the means of a permanent Water or strong Vinegar, which is the quintessence of the matter and Soul of the Stone, but let us note as we go that the metals do participate of this radical humidity, the which is no other thing then the beginning of all other soft things. Therefore it is an assured maxime that the Calcination of Philosophers is no other thing then Whitening and the purgation, restauration of natural heat and radical humidity and the only means to expel the Superfluous humidity, and an attraction of fiery moisture which is this pure white which we call the internal SULPHUR of Philosophy separating from itself all accidental and superfluous Sulphur which is Corruption; Otherwise a pleasant Liquor from which proceeds the animated substance of our work, the sovereign quintessence of all happiness the chiefest Spirit and the life of which is extracted the compleate redness. The glorious Crown of our Labours, not this liquid Substance is ordinarily made with the Water of the Philosophers which properly is the Solution and Sublimation of the Sages, or their Exaltation or Whitening of their permanan.t Water. And of such particular force that it suddenly changes the hard dryness into a supple and manuable drawing out the Quintessence which is the admirable Stone of the Wise men and the Vegetable Mercury which separates and conjoins the

Elements which happens principally because the parts
with the violence of the fire hath consumed and
composed together is become subtil by the Spirit
which is a resolving Water and a humidity of
uncorrupted bodies gathered together and annexed by
heat, to the Spirit and radical humid. All which
thing makes one root of the philosophical Elements,
the which we must renew after Corruption, which are
the four perfect Colours, the red being last
according to the fountain of the Lovers of Science.

Then guide by reason, thy intents

To divide the four Elements

All which thou again new shalt make

And them into thy work then take.

Now the Sublimation is named a terrene vapour gross
but subtilized and brought into a humid Vapour, or
moist Air, by the well-tempered heat of fire which
heat absolutely causes the motion and necessary
mutation of the Elements, which whosoever knows
these mutual Conversions of one into the other, may
rest assured of his Way, whereby he may find the
Quintessence extracted out of the entire Elements,
not any more to be combured with superfluous
humidity or foul pollutions.

But this Quintessence is an opaline humidity of
excellent quality the which gives Luster to the four
Elements, without restraint, transmuting into the
proper nature of a quintessence and then called the

Soul of the World and fire of Philosophers. This is
even the true fixation that Geber mentions; Nothing
(saith he) shall be made firm, but by receipt of
some Light, or when it becometh a fair and
penetrating Substance, for of that cometh the
Sulphur and Center of the Philosophers, which cannot
be extracted without the Lune which is the chief
point of their Mastery and their greatest Secret for
in it is the metalline Water preserved, the which he
receiveth of the body he hath animated and restored
to life, this is a mixture of the white and red
tincture and formative Spirit.

For the moon doth covertly contain in it the
tincture of the Sun, the which he doth produce in
the form of a red Sulphur, at the End of the
Decoction, all by the means of the Soul of the World
and the fire of the Philosophers which doth all of
himself.

Yet in this ablution much blackness and Corruption
doth appear through the heat of the fire which doth
putrefy everything, and Whiteneth the black things
which once were dead and brought to naught at the
same time restoring life to the matter in the which
one may perceive a pure and intire heat intermingled
with a kindly metallick humidity from which the
matter doth receive tincture, virtue and vigour. The
putrefaction so much desired of all the philosophers
which is their choice Study shall be perfected and
accomplished when it shall manifestly change and
alter its first form and from a black colour become
a white, the Secret being produced by Corruption for
that, that was hid doth show itself apparently to

sight and reneweth itself from death. Therefore ought one to have especial heed in our Work to the black essence of the Sulphur of the Philosophers; This is the same that Arnold de villa Nova saith in his Rosary: The perfection of this Work is in the change of natures; Of the same opinion is Raymond in the Theorick of his Testament, The Art (saith he) of our Mastery, dependeth on Corruption, and we dissolve putting it to putrefaction at the same time, and elsewhere he saith whosoever doth know the means to destroy, that is to dissolve the Gold he hath attained the Secret, and our Stone is not found but in the bowels of Corruption. The Turba of the Philosophers saith further that Corruption is the Ascendent, and chiefest hope of all the Work, the which doth discover, and manifest the highest Mystery of this operation, which is principally a certain destruction and true Conversion of the Elements.

That which is manifest we hide from sight

And bring the hidden thing again to light.

It is of this change that the learned Turba gives us so often admonition, Saying change the Elements and make that which is moist, dry and firm, who yet passes further, answering as that the matter with all his dependence is then prepared as it ought, when all is pulverized and brought into one body, Which for this effect is most properly called of the Philosophers Conjunction. Consider also and pray you, that your Calcination, is in vain; if thereby no powder be produced, which is the water of the

Philosophers. The Ashes of HERMES or powder of Mercury as AUGURELL saith in these terms:

The Water which I mean seems to the Eye

A powder and is so called properly.

Decoction is also one of the principal and necessary parts of the business whereof they ought to understand the Mystery, the means to employ the flower of their best Decoctions in the Essence of our Mastery. Albertus Magnus concours with all the other philosophers in the same opinion. All holding it in special esteem.

But seeing he is the first that presents himself, I will for the present repeat his speech. "Of all the Arts (saith he) yea of the most perfect we know not one, that so neetly imitates nature as that of the Alchymist, in the decoction and formation of this red and fiery Water of Metals attracting the vital qualities of the Sun, and so small a nature, also the Philosophers have an Assation and usefull dissolution, by which the humidity shall consume itself by little and little and through the fire become a body more powerful against the flame then metal; But we in our Work must take good heed that the Spirit be not too much scorched and dryed so that he may no more hold correspondence with the Body, and not be sufficiently purified and perfected."

Distillation of the Philosophers otherwise called
Clarification doth also bring a great advancement to
the Conclusion of this Work, which we hold to be a
Plain purification of the matter with the radical
humidity, the which being found gives hope to the
Sages of a desired rest to their almost tired
spirits, by means of this Coagulation, the perfect
alliance is made and the Corruption of the Sulphur
not Vulgar and the Crow or Bird of Hermes which
always flyes about the Tops of the Mountains that is
to say upon the superficies of the metals, with it
is a spirit black and not burning insensibly Crying;
I am the White of the Black; and the Red of the
Citrine, I have with a pleasant riddle expressed
this bird, which I will here set down, finding it
very suitable to our subject; In memory whereof it
was Learnedly composed the modest Curiosity of our
Mystical Work being thereon comprized.

The Riddle.

I dwell in Mountain Tops, in Valleys and in Plains

Father before a Son, my Mother was my Child,

Born in my Mothers womb, my Father first exilded

There without nourishment, I did myself sustain

Hermaphrodite I am, both sexes I maintaine

The strong I vanquish yet am by the weak beguiled,

There's nothing under heaven then I myself more
vilde,

Nor ought so fair so good doth in this World remain,

In me, without me strangely one strange Bird is hatch.

Which of his bones, no bones, builds for himself a Tomb.

There without Wings he flies for swiftness never matched.

By nature and Arts Law received hath his Doom.

In fine he yet revives, and makes himself a King.

And to his breathren Six, he golden Crowns doth bring.

The Rosary speaking of this Coagulation compares it to the CROW that flies without Wings, that which is principally done by dissolution caused with heat and Congealation the effect of Cold which two, are the means of perfect Generation.

Hermes speaking with what kind of heat, the whole Work should be governed saith in his Smaragdine Table that the Sun is the Father and the Moon is the Mother and the fire the Governour saying:

All perfect and intire is then his force

When to earth again he hath recourse.

And when by degrees this Elixir comes, to settle in firm earth the which afterwards may serve for so many several operations as cannot be numbered upon any apt body, to which it shall be applied, for which reason we may compare it to a commodious Gardner which safely preserves all manner of Grains to use and profit. So our Art once perfected converteth all things that have propinquity fitting into his Viens Excellency of nature and being furnished with sufficient materials, raises admirable Structures, resembling the perfect Architecture of the Sun.

Of the diverse Operations, the Various names, frequent in the discourses of this Art.

It is a general saying amongst the philosophers, that he that can kill and fix the Volatile essence of the fugitive Mercury, shall attain the excellent operation of the metals, and know the greatest Mysteries of this Art; Yet must we not premptorily pitch upon the rude Letter but seek out some ingenious glosse that may discover their sense and meaning; because they speak diversely of their Mercury and here we think fit to place in the frontispiece of this their innumerable controversities a Sentence of Senior for the preeminence he hath before other Authors; Our fire saith he is a Water. But when you can fit one fire to another Fire, and one Mercury to another Mercury. This knowledge shall suffice to bring you to the glorious End of all thy Intentions. Here note that this Argent Vive is called a fire, and a Water and yet it is necessary that this fire should be made by

means of another fire, he says also in another place
that the Soul must be drawn out by Corruption, which
is blackness and the first Colour of the perfect
Elixir the which infuses itself again into the dead
body to preticipate his Spirit to it, and to give it
a life and resuscitation and to the end that the
wise philosopher may afterwards peaceably enjoy both
the body and the Spirit by this perfect operation.
It is the same also that is spoken in the Turba
where they call their Mercury their Fire. Take
(saith he) the black Spirit not burning with the
which thou must dissolve and divide the Body. This
Spirit is all fire dissolving all sorts of bodies by
his fiery property. Others hold that this Mercury is
properly named a Quintessence, Soul of the World,
Spirit, Water permanent, menstrae and an infinite of
other names. All which they impose upon him
according to the diversity of his several efforts to
whom they attribute so much power and virtue as that
without the assistance of this quickening Soul, the
body of our Vessel which is the black matter called
the dragon devouring his Tail, which is properly
humidity, should now regain life nor demonstrate any
good effect. Take say they their Quicksilver and the
Body of the black Magnesia or some pure Sulphur not
burned, which you must pulverize and grind in most
strong Vinegar; But you shall not find any apparent
Change or mutation in the Colours permanent, the
black, the white and the red which are the most
necessary. If the fire do not make White, nor
approach this Composition. For he only takes unto
himself this property and indures the perfect
regiment tinging him into a perfect red, throughout.
The Turba say it will become Gold and transplanting
itself into an Elixir, from whence one may extract a

Water serving to diverse tinctures giving life and Colour to all that are Joined with him.

For as Blackness is the first that displays himself in the Work, so doth it direct the assured march of the rest, and as it doth precede all the others, so is it the foundation containing all the rest potentially as Arnold affirms that whatsoever Colour appears after black is Laudable for whenever thou shalt see thy matter turn black, rejoice and comfort thy self because this beginning shall continue assuredly to a happy End of the whole Work; in the great Rosary it is also said that all the perfection of this Science consists in the transmutations of Natures which cannot be attained without passing over the black Stigian Lake described by the philosophical poet OVID; otherwise you are out and must begin again; though never so much against thy will. But if you can perceive in your Vessel, the black Sulphur, whereof we now intreate, It is a perfect and infallible Entrance to all the necessary ways. See thou the great Esteem that the grave and provident Turba have of this Original Colour which doth precede the Citrine and the red Colour outwardly appearing praised and exceedingly hopeful and fairly promises good success, after which comes a purple very precious and of great comfort making assured the happy Event of the triumph and magnificence praised to our king, and this Colour is the best and most pure Mercury which furnishes us with the most exquisite tinctures of our Mastery, endued with a most Sweet odour. Now all these beautiful and excellent properties, attributed to this worthy Mercury do manifestly show the cementing power and the subtle Vivacity of this volatile Spirit.

Hermes that great Prince of Philosophers, ignorant of nothing that was natural, had his Spirit so transported with the Contemplation of the Excellency of this Mercury that he professed himself unable to give this mineral an Epithet fitting his powerful and glorious effects, yet willing to deliver a metaphorical abridgement of the particular properties thereof than describe it.

Most common, most Unknown, most precious and most vile.

Conserving and destroying, both you may him stile.

Good and milacious, beginning and the End.

Of Treasures nowe afar and nowe again a friend.

For Corruption and blackness are the beginning and end of all things, and Augurellus in his Crysopea affirms as much of that black bird that dissolves all bodies in this Verse:

And which is more, this bird so potent is

That he dissolves the metals without miss

And naturally is in every thing

First in their birth and at their last ending.

The Axioms and principals natural assuring us that
Universal Corruption is the common Sperm and proper
seed of all Generations. But in fine to return to
the nature of this bird in whom we may mark and
perceive such a power that he is able to withstand
whatsoever is contrary to him taking his flight
sometimes to the Sign of Leo, sometimes to Cancer
and other whiles to Capricorn, but if after so many
subtile flights thou canst stay, qualify and correct
his fickleness stopping the Swiftness of his course,
thou maist purchase the precious Golden Loadstone of
the most rich minerals, and thou maist at length
enjoy many precious things whose exquisite value now
came within thy Imagination.

And then thou must separate and divide him into
diverse parts reserving to thy self always some part
which thou shalt again reduce to his putrefied and
dead Earth so long till his volatile Spirit lend his
aid to set him upon his foot by his natural
Strength; beautifying him with Variety of fair and
pleasant Colours, most plain Evidence of his
Clarification. And when all this is past it is
called by all good Authors the Earth and Lead of the
Philosophers which they may happily make use of
having now attained the quality of heating the
Vessel of Hermes, which is Mercury and when and how
to distill by number or plain Distribution,
qualifying this Spiritualized Earth with diversity
named according the Successive Colours and diverse
operations of this wingless flying Spirit subliming
and rectifying even to the bottom of the mass, which
decreases, purifies and renders itself more and more
fair in tincture even to the perfection of the first

White which must again be mortified, and afterward
restored to a more glorious life, which is the red
tincture; Putrefy again this body and pulverize it,
till the occult and the more hidden red come forth,
and be manifested to sight. After this dissolve the
Elements and separate them in such sort as thou
mayst again rejoin and reunite them according to the
manner and again putrefy so oft until thou hast
brought the corporeal and material substance to an
animated and spiritual Essence, which being happily
done, you must again draw the Soul from the body
which you shall again rectify with his Spirit.

This gentle Messenger of the Gods Mercury full of
invention and subtility, being thus often
metamorphized hath gotten to himself much Luster
whereof he makes large and liberal portions to his
Associates and nearest neighbours; As to Venus on
whom he bestows a rich white; he moderates the
Crackling of Jupiter bringing him to Solidity;
hardens, whitens and fixes Saturn; Softens Mars
making him fluxible; gives unto the Moon a glorious
Citrine Colour and resolves into a perfect water
from whence may be extracted an Essence of admirable
Virtue. Trevisan openly delivers in his practick of
the natural Philosophy of Metals, to which we refer
the diligent Reader, the philosophers do point us
out with their finger the necessary means to attain
the preparation of the black Sulphur, even to the
first nature of red which they call distillation,
until it comes to an oleaginous gum and waterish
incombustible, very penetrating and altogether like
the body.

Wherefore it is called by many the Soul, because it revives, conjoins, renders and reduces the natures into Spirit. This Sulphur thus reduced transcends in excellency all the value that can be Imagined or expressed and therefore have they highly praised it and given it a title of great honour attributing the prerogative to it of the rare name of Lac Virginis which returns in some sort to the form of a red Gum all of gold resembling the Water of the philosophers most resplendent, which ought to be coagulated, commonly called of the Sages TINCTURE SAPIENTIAE, the admirable tincture of Wisdom or the Vital Fire of the permanent colours. A Soul, a Spirit that by his virtue much more altereth himself at his pleasure becometh volatile and contracting himself when he pleaseth, of a high fixed tincture in his individual, that is his own proper homogeneal nature.

This Mercury not common is yet called red Sulphur, Gum of Gold, apparent Gold, the desired body, most precious Gold, Water of Wisdom, earth of Silver, White Earth, Air of Wisdom (note that the Child of the Philosophers is born in the Air) then principally when he is become very clear and perfect white.

All the Turba treating on these Circumstances that do appear upon the Surface and upon the entire body, of their fruit have given this Judgment: He ought say they, to know that one cannot tinct gold into red who have not passed the white after Corruption, because there is no way between the two Extremes of the Work but through the White Work that is the middest. Therefore ought we intirely to observe these methodical rules seeing the discord and Center of Confection which he doth ever hastily run into through the way of desolution who hath over run the good Concord directed by the prudent discipline of a well advised order necessary to this work. Now all these Colours aforesaid are of the same nature and are successively found in the same subject, though they produce divers effects. For it is truth that the white shall be made black by the red, and that from a pure Crystalline coloured Water, there shall citrine red appear altogether, several of the said philosophers secret virtues.

Morien treating metaphorically of the transmutation of metals, of the proportion and degrees which ought to be observed in the Composition of thy work. Cause saith he the red fume to comprehend your White fume, Also pour them down to the bottom and there conjoin them together.

The Codex of all truth saith to the same purpose, blanch the red and make the red white and then hast the whole art from the beginning to the End. Senior also speaking of the Varieties of Colours, gives us to understand in the Words following the great profit and necessity of them. It is an admirable thing to consider the wonderful operations and noble actions of this mercurial Spirit. The which if thou project upon the three imperfect planets he makes them rich in whiteness and upon the other, as rich in redness and Citrination, the first whites then the Lilly or the untouched snow, the second more orientally red then the poppy or the ruby. To which form Morien confirms his judgment though in other Words, and by another Way. Take heed saith he to the perfect Citrine which by little and little divests himself of his Citrinity to gain a more glorious robe of unspeakable redness after the dismission of a formed blackness, strong and powerful, which she was fain to wear in her younger years, that she might so serve as an earth, a base and assured foundation to build the rest of the work upon.

By all these inviolable Theorems fixed to the Ideas of the most famous Architect (which have happily undertaken the industrious fabric of this excellent Stone, and framed with an Artificial hand the true Cube of Hermes) we may easily conceive that the Gold of the philosophers is absolutely another thing than common Gold, or silver which yet are indeed of the nearest neighbourhood unto it, and the chief Imitator of their Golds perfection. For although the Similitudes (which the Sages the Sons of this Science) put between them make as though there should be community and familiar Conjunction with Vulgar Gold and Silver, as also with other metals which fail in Effect the purity and perfection, therefore we must ingeniously consider that they induce such things but by way of Composition meerly.

For the profound Raymond most charitably tells us once for all that out of metals perfect or imperfect we have no need to extract a Spirit when nature herself as our handmaid hath prepared one for us which Spirit (saith he) we find only in our own (not the vulgar metals which are dead) from whence we artificially draw him, and again conjoin him to his body that so we may be masters of the vegetable Mercury of the Philosophers; An Axiome worthy to be ingraven in gold. And for common metals notwithstanding that the many Authors are of opinion that the impure metals do ever remain such without reaching to any higher Luster and that Lead always retains the nature of Lead, yet we may separate for some special property of excellence even in these; and the reciprocal Affinity between them, and the Elixir that works upon them, they need not be assaulted of the Comparisons they so often use in this kind, they being (if well understood) so full of expression and demonstration. Consider that which much to this purpose Senior reports speaking of the imperfect, which notwithstanding saith he pretended one day to be equal to the most perfect, which no way exceedeth them in nobleness of Essence but in primogeniture, only having had a longer decoction, their extraction being as Vile and abject in the natural Composition as the imperfect, the most perfect of them being originally without difference of nobleness in the common Seed and universal principle of the most abstract sordid metals. I am (saith he) speaking philosophically) more than metallick Iron hard and dry but such is my power and virtue that nothing may compare with me, for I am the Coagulation of the Argent Vive of the Philosophers. The Turba also says that prepared lead shall become a Precious Stone, qualifying the most noble and perfect Colour of the Work yea, the work itself which they name Copper.

They say also that Lead is the beginning of the true Mastery and the thing without which nothing can be done: They have expressed as much of the red lead made white, or Venus of Mars: And of white Lead (as they continue their discourse) thou shalt make white tincture, which is the Lunary Sulphur and then shall thy Labour have passed the blackness; and hath arrived at the White, the Second Lady of our kings Officers, and the proportional middle point of our Artifice. And for this cause the philosophers have taught us that there is nothing of nearer neighborhood or that doth more approach the nature of Gold then Lead, for as much as in him consists the Life which attracteth to itself all the Secrets.

But we must not take these things only literally, nor seek in common Lead these rare phenomenon, in whom these properties are not to be found, except only in that which is called the Lead of Philosophers as well for his facultie in putrefying as the infection of his stinking earth, he is advanced above other metals.

This is the reason that they all conclude with Raymond Lully that without putrefaction the Work can never be effected, it being the Water, the Fire and the absolute key of the perfect Magnesia.

And to this purpose hath Morien learnedly compared it to Arsenick, to Orpiment, to Tutia, to rotten Earth, to stinking Sulphur, to all kinds of Venom, poison, and Corruption for the Correspondence that it hath in some quality or other with all these things.

And further to diverse bodies which are not of the number or nature of minerals, but that only retains a Commerce in Complexion as blood, hair, eggs, and many others. And finally to diverse Mineral matters as Salt, Allom, and infinite others, in all three Regions, Mineral Vegetable and Animal. All these varieties being Attributes, in regard of the apparent Conformity that it holds in effect with every particular Genius and Species of those bodies and Spirits aforenamed.

For which Cause Geber affirms that their Stone is extracted from the metalline bodys prepared with their Arsenick, that is to say their Corruption, and Calid in his Secrets saith; "Annihalite the leaf with Venom, therein denoting putrefaction".

But above all Alphidius advertises us to take great heed in the wise Government of an animated body or a mortified Stone which is the blackness. For (saith he) as by the privation of their Natural heat the which decays even to the death, being now destitute of all his first functions. So if for remedy thou thinkest to give them a greater heat then is fitting to hinder the perishing of the heat, with which they were naturally entertained and nourished in Corruption, thy matter shall become red before black, which is the privation of life and thou shall loose thy Cost and Labour. For which cause we must accommodate ourselves, with a most gentle fire and naturally well-disposed to the end to revive that which this privation hath debilitated, by his offensive Violence, for as Ripley saith in his 5th. Gate.

Thou keep in temperate heat eschewing evermore that they by violent heat be not incinerate to powder dry, unprofitably rubification; but into powder black as a Crows bill with heat of Balne or else of our Dunghill.

Above all things remember to keep them in a moist heat until fourscore nights be past, and that the black Colour appear in the Vessel, which is the first Salt of the philosophers and a tincture near the quality of Sal Alcaly and other Salts, of bodies that which changing itself subtily into the nature of the things attracted becoming all one with the natural Essence of the metallick nature.

Now the philosophers do diversly handle these varieties as well of their Stones as of their Salts. For as much as the greater part of them do constitute three Sorts in the perfection of the entire Work. And for witness and warrant of this Thesis I will take the proposition described in the Great Rosary in this manner. There are three stones and three Salts out of which the whole Mastery existeth. Lucas Rodargirius in his book of the philosophical dissolution, where he makes an ample discourse thereof, rests himself resolutely upon this ternary number.

But we must not forget Raymond Lully that calls these three Salts, three menstrues, three Vessels, three quicksilvers, three Sulphurs, three fires, which are no other thing (to speak properly and not hyperbolically in dark philosophy) then the three Colours, black, white and red, which are only extracted from the natural essence of the true matter. The which Salts have so much power upon the perfect beings of our Mastery that Senior speaks thereof in these terms, our body shall first become Ashes, afterwards a Salt, in fine it shall arrive by his various operations, to the measure and most perfect degree of the Mercury of the Philosophers.

But amongst all the Salts, it is to be noted for the total instruction and fabrick of the Work that the Armoniack holds the prime and principal place surpassing in Excellency the impurity and Essence of the others not so noble, which for this purpose of our Work are found also prepared by many degrees. As Aristotle himself assures us in diverse particulars of his work advising us by his discreet pen only to use this Salt Armoniack in our operations because that only hath the property to open, dissolve, soften, and animate the bodys. Now there is nothing animated or engendered without a precedent Corruption, as Morien says which is the black Colour or this Sal Armoniack and the black Spirits dissolving the bodys.

The Turba adds abundance of these Speeches to confirm our Affirmative. We must (saith he) know and perfectly understand that the bodies will take no tincture unless the Spirit hidden in their Bellies which is this black Spirit, be not from thence with great Dexterity and difficulty extracted, which being done (as it ought) there shall come a water and a body that resembles our human nature, for it then contains body, Soul, and Spirit, the which while it is an essence of a mean Colour cannot perfectly tinge the gross Terrestrial substance, if it be not subtilized, by this Spirit and made like unto it and the Spirit of Waterish nature be tincted into Elixir, producing white and red with pure and perfect fixation, high in Colour and of a penetrating tincture mingling itself indifferently with all the metals, as the Celestial Mercury rejoining himself to every Planet becometh of their nature, be they noble or imperfect.

But yet we must know that the perfection of all the mastery depends upon this only point that we draw the Sulphur out of the perfect body having a fixed nature for this Sulphur is the most ancient and most subtil part of the Crystalline Salt, odiferous to the smell, delectable to the taste, and of an Aromatick humidity, the which being the Space of a year, in the fire shall ever stand as melted wax and because he holds some part of the nature of Quicksilver he tincteth them into a most pure Gold and being the water or humidity which is drawn from the body of metals it is called the Soul of the Stone and that which is hidden in this humidity is called Spirit and the virtue of the Spirit is called Soul and tincture which tinges and fixes all the Water into pure Gold.

But the Mercury or his force and vigour is also called Spirit when he hath attracted to himself the Sulphureous nature. And the dry earth is the Body and the body is of the Quintessence and the extreme and absolute tincture, which is the true Essence and perfect nature capable of all forms.

Now though these three proceed from one root yet hath they notwithstanding very different operations, the names of which are infinite according to the apparent Colours and if all be again reduced to one, to wit to this final redness, it serving as Links so artificially chained together yet it is a great difficulty to discern an absolute end, for the one finishing his ordinary Action the other recommences anew, because according to Raymond; the first form being destroyed there is another immediately introduced, and in his Testament he calleth it the Golden Chain which doth link together the visible to the invisible, uniting together in an indissoluble bond the four Elements.

John Mehurig saith in his complaint of nature:

This is the most rich gotten chain

Which circulerly maintaine.

And in his Romance of the Rose he calleth this Mercury a whore which conjoins herself indifferently to all forms one after another.

The admirable Virtues and more than humane power of this noble Tincture; briefly and perspicuously declared in this our Instruction.

The exacter the tinctures are, the more acceptable are they according to the usual fashion, that bears rule and sway amongst men through a desire not unworthy but rather most commendable in Ingenious Spirits curious of the inestimable Value of any rare novelty so well for the benefits that doth almost equal this Curiosity, as for the desired honest and prerogative befitting their noble disposition happily, at last attained through the absolute possession of this pleasant fruitful of felicity.

This is the directest and most apparent means to persuade even the Soul with Sweet hope and with a calm Gale of a pleasant Aire and very seasonable to satisfy the most earnest wishes by the gain and full contentment of aforesaid proposed object within the Idea of our fancies premeditated before the happy effecting and fruitation of this Delightfull possession. Seeing that naturally we wooeth (reach) after things which are amiable deservedly believed and desired, for the cause principally heretofore mentioned, of greater reason ought we desire the Enjoyment of our marvelous tincture. But because we can hardly endure the pain-full search of an unknown thing principally seeing that the real and actual knowledge ought first to be apprehended within the subtil veritie of a quick apprehension, which he may potentially attain, and be assured of by the forms sincere, friendly descriptions and that the general intention do first aim to know the lovely thing

before it be beloved; I have handled in brief Words
according to our fashion, the intricate Course of
the pleasant operation of our natural Science;
issued and drawn from the pure and perfect testimony
of the Ancient Sages, which I may call the chief
Judges gratiously advanced to such Authority by the
Supreme License of the Divinity and by the Sacred
Conceptions of the mysterious tree which they have
admired for his Sovereign balm, to the end that by
the true knowledge of this rare virtues and
particular qualities, every virtuous Soul persuaded
by sound reasons, grounded upon the excellent Luster
of this glorious tincture suddenly yielding their
Spirits amorously surprized with great admiration
lay hold of the Skirts, of some benevolent virtuous
man, as the ordinary gages of their fidelity and to
announce to all the Sages, the esteem they have of
this same excellent subject altogether venerable
most transparent to the Eyes and by his sweet odour
better apprehended to be of such a harmony. The
delight of which changeth, the over whelming waves
of so doubtful a shipwreck subject to the mercy of
many bearfull Irresolutions as a small boat is
directed by the safe proposed Sea Snake and the help
of the nautick needle, rather maintain the vessel at
the End arriveth happily in the safe port of Comfort
with joyful sails through the skillful Conduct of
famous pilots and by the ensign HALCYONS of the
Jasonicke Hands, who shall do this, Let their hearts
be even ravished at the Sepulcher of some Saint
Anthony. Let them be firmly delighted in the sweet
Register of such a remembrance, Let them perfume the
alters of their ardent devotion within the temple of
honour and knowledge, with some being art of pious
humility in sign of their compleat Joy and Ecstasy
of heavenly Contentment surpassing the superficial

appearance of human contemplation, of which the grave Ideas only are in possibility to ascend the Supreme top of the most lofty mountains approaching heaven. In their intelligence of formed Essences by the Lively Effigies and natural representation of a terrestrial Sun shining here below as well as the Celestial after the same sparkling brightness somewhat illumination mens hearts making them more Zealously to acknowledge the Soverign duty they owe to him manifesting their ardent affections by the sweat of earning bowells of universal Atoms of the image of his Glory within the delightsome Angles of the terrestrial minerals by the profound inspection and sublime preparation of a mysterious philosophical and most admirable Art.

I will speak now of our tincture wherewith the animated spirit is after a sort made perfect which doth entirely perfect the most perfect Colours.

Naught else like him is found

Of his alloy so sound

In his proper Essence

By sole Actioitie,

Surpassing happily

The purest Excellences.

The ancient Sages prudently observed four remarkable points in this vital power extracted from the great

number of his proper virtues when that the
properties are fortified by infallible maximes which
the same nature shewing disdain seems almost
discontented by the difficulty of his assurance for
the approbation of obtaining so great qualities.

By agreement free and voluntary

This power is all in all ordinary.

It is true the greater part of his Virtues are more
than can be Imagined, esteemed of some as a thing
impossible and contrary to natural reason, which
gross, Ignorant Dulpates will not willingly
acknowledge that any other hath that which they
understand not with Jelious vaunting as though they
could fathome the depth of these more than human
perfections and determine of so great prerogatives
by the rash Sentence of a shallow Incrudulity.

The fond conceite of vain appearance

Of chanced act, without Experience

Demonstrated, shows but presumption.

Thus not stretching their Conceits further then
these biased fancy's filled with vain Scruples and
with a more then panick and apprehend Errors or
great Contempt of our Mystery. But what have I said,
yea are they not much rather taken with a confused
fanatick Censure of extreame shallow brains hammered

on the unpolished Anvil of an absolute carping Zoylus.

A discovery of the marvelous Effects of the true medicine of the Philosophers reduced into four especial and remarkable points.

The first point of his perfection is to preserve the person of man in his entire ability and strength free from any accidental malady that may assault him and to confer upon him a perfect Constitution and healthful, and vigorous disposition with a merry Spirit till he may be a competant member of his posterity chasing absolutely by the Virtue of his operation the threatening causes of our evils which otherwise would daily corrupt and overwhelm us with frail Infirmities without the aid and help of this sovereign Antidote. Calid in his mirror of the Secrets of Alchemy writes that it mundifies the body of all diseases and conserves the well-tempered Substance and the Vitals in their entire prosperitie exempt from any Imperfect alteration.

The Second accomplishes and makes perfect the bodys of metals according to the Colour of the Medicine, which if it be white it transmutes them all into fine Silver; If red into most perfect Gold.

The Third changes all sorts of Stones into precious Stones, after the measure of the decoction that the medicine hath gotten.

The Fourth works upon Glass reducing it also to the nature of precious Stones of which Colour you desire the medicine having been first decocted more or less to the purpose as shall appear by the following paraphrases upon these four points.

The Mystical Work of our Stone being perfect and wholly complete is a gift of God so precious that it surpasses all the marvels and most admired Secrets of all the Sciences in this world for which cause we call it (in imitation of many excellent Authors) the incomparable treasure of Treasures.

Plato hath so highly prized it, that whosoever hath acquired (saith he) this Gift of heaven he holds the best of this world in his possession being mounted to the height of riches, and the most sovereign physick.

The Philosophers ascribe unto it the virtue of healing all sorts of persons detained in the Languishment of any disease whatsoever, by taking in drink a little warmed or mingled with wine or in the Water extracted from any simple that hath Sympathy with the part offended or infected receiving in one day a disease of a month; or continuance in Twelve days that of a year and in one month the most invenerate and Chronick.

The dose being no more than the weight of a grain, a greater quantity not to be given without the prejudice of the party, Let the mallady be, dropsy, Gout, Leprosie, Apoplexy, Collick, Headache, Frensie trembling of the heart, fever, falling sickness, defluctions of all sorts inward and outward this medicine makes a quick hearing, and it fortifies the heart, recoborates the imperfect members, chasing out of the body all fistilas, Ulcers, and imposthumes; And in fine it is the true balm for any ill, and a singular preservative against all corporal infirmities, poison, or other reviving the Spirit, augmenting the strength, conserving youth, prolonging old age and chasing away evil Spirits; And so qualifying the temperatures that no humour gets predominance of others, to alter the complexion and condition of the body otherwise then for the bettering thereof.

Briefly in this Work is fully seen the great Secret and Incomparable Treasure of the most rare mysteries of all the Philosophers. Which Senior confirms, saying that this projection renews a mans youth, renders him Joyfull and merry, conserves him in health, to the end of ten ages.

Our famous Roger Bacon affirmes Artephious to have Lived by virtue of this medicine 1025 years.

Wherefore not without great reason Hypocrates, Avicen, Galen, Constantine, Alexander and many other physicans (whose memory the World celebrates) prefer the Elixir before all their medicaments, who so

terming it the most perfect and absolute medicine
and the universal balm of the World.

For the Second it is held for an undoubted maxim by
the experience of divers authors without all compass
of extremes that it transmutes the Imperfect metals
in a moment into pure Silver and Gold most perfect,
and far exceeding in Colour, Weight and substance
and constancy of trial all mineral GOld and Silver
whatsoever and so high in value as no refiner in the
world can make a just report of his Carrats.

For the third it is most certain that this powder by
projection upon other common Stones (being first
liguified) doth make and produce most precious
Stones as Jaspers, Amethyst, Hyacynthes, Topaz,
Chrysoliphs, Saphires, Emeralds, Rubys, Diamonds and
flaming Carbuncles, much better and far excelling in
Luster and virtue, those that nature doth produce;
all the which this medicine can liquifie.

And for the fourth and last property of our Mastery
it hath the virtue to communicate itself to
vegetables and Animals and to every inferiour body
to make them perfect; Yea there is not the most
Simple creeping Creature in or upon the face of the
Earth that serves not as a sounding trumpet to
announce the glory of his excellent prize. Even of
that which if you project a little upon molten glass
you make it malliable and of what Colours you please
and as he proceeds to his purification in his
decoction when he is green you may make Emeralds
when he is like the rainbow (which appeareth in the

vessel before the White) he makes opals in his Ash Colour he produces Diamonds, and in his red Rubies, and the invaluable Carbuncle. But for fear the Sages should envy my pen for having so punctually and perhaps too boldly pointed upon this Table to the open view the Secrets which they have with much care and Cunning shadowed with so many obscurities and vailed under so many Hieroglyphical figures as it cannot be discovered but by the ripe judgment of some prudent Oedipus.

I will here make an End excusing that fault, with the feeling I have of the sufferings patiently borne by the virtuous Children of Art which this World makes miserable whom I confess I have a Will to help. And for the sottish Ignorant I now doubt their approaches, for where there hath been so much Art used in hiding as they that are of the notion and understand the Language failed in the finding, it is not for Idiots, nor vain glorious Empty Thrasos to hope it being to them a Gordian knot which whosoever will untye must be furnished with knowledge, patience, diligence and virtue.

Now upon whom should so great a blessing be conferred if not on those Eglets which ran with open Eyes behold the glorious Sun of Philosophy (that strikes blind all others that unworthily stare upon it) and for you and their deliverance out of the Iron-fetters of Contempt and misery, I have written this plain and true discovery of a kingdom that far transcends Arabia foolix or the East and West Indies happily and peaceably may they find it and injoy it: that we may so Plato's plot accomplished which was

to produce a Government wherein the King should be
Philosophers and Philosophers kings; and not as
nowadays when fools ride on horseback in state and
pomp and wise men larky by them on foot by them, as
despised Attendants.

The Conclusion.

The Work most perfect, most commendable and most in request is that which brings to the workmaster the fruition of whatsoever he can wish for his Commodity or pleasure and which defends him from all the importune streaks of Judgment, the publick plague and conspiring Enemy of all humanity and especially to the best Spirits the worst Tyrants.

Now if by the powerful Antidote of this murdering poison, man may dissipate and happily blow away the noisome vapours of his sufferings to savour and taste all leisure the comfortable and profitable fruits of his industrious and diligent hand and ingenious Spirit that is desirous to give testimony of his good Will and Charity towards the necessity of his Companion and by some Charitable Art, to relieve him and release him out of the loathsome dungeon of careful necessity.

No man that hath any sense of Virtue, seeing the Effects of so wonderful a workmanship can forbear to admire the Authors Love and honesty the Art that thus doth conserve the comfortable society of mans life, in a fair, free and flourishing estate.

Shall we then remain bruitish without consecrating the famous Sacrifice from the Altar of our hearts, to the lively memory of our admirable tincture which without Comparison places the professor above all other men, advancing him to the highest stop of human felicity. Shall we now in this happiness become Stupid and insensible of the honour due to so sublime a business, seeing the unfitting and too ungrateful Silence of mouths indiscreetly mute would have but in this respect but small Grace; If we cannot excuse this fault with the dread we have that we cannot sufficiently extoll the subject furnishing so ample a matter of discourse.

In such a Case insufficience might have a place in our writings. Whereas a rude neglect in not acknowledging of such an Artifice so absolutely excellent, as nothing in this sublunary vail can equal were gross and absurd in the Eyes of the judicious who could not but condemn with a publick Curse, such, as should by contemplation or deny all, blaspheme against the true Essence, and real nature of this inestimable work of philosophy.

The most perfect divine Image where,

The glass wherein all natures works appear,

Giving us whatsoever we hold dear.

But although the Work of itself be plain conspicuous and easy the base being true SIMPLICITY BEING THE SEAL OF TRUTH, yet because it should not be prophaned nor so precious a pearl cast befire Swine the best advised Philosophers have treated thereof in figures, riddles, obscure parables, Circumlocutions, hyperbolical Dialogues, and shadowed Similitudes, to the End it should not be contaminated by impure and unsanctified hearts and hands of abject and vile persons as is requisite in so Sacred a Mystery.

Cease therefore pusillanimous souls, Cease to sweat any more in vain undertaking to beat out the path of virtue. It is to you of difficult Access and full of hazard and ruin; but to generous, patient, painful and ingenious Spirits the new soul at the apprehension of so dangerous a passage having health, wealth and a mature apprehension of their proposed Labours. It is but Childrens Sport and Womans Work, honour also taking pleasure in this their bold and brave attempt conducts them by the hand through all crosses and never forsakes them till they be arrived at their desire and felicity, and triumph happily in the plentiful harvest of the seed sown in the fertile Globe of their perseverance which grows in the End to the palm of glorious victory. The valorous Argonauts could not be diverted from their Enterprize by the pillous Sands nor threatening rocks, but made all good into the point of their Constancy and gathered the sweet fruits of their expected glory, which a timerous Soul durst never endeavor to attain nor expose his Sail to the Wind in the violence of unmindful Waves for the honourable Spoil of so rich a bootie. So may we say of our Work whose navagators are selected and elected by the Councel of Heaven neither s-bricking Sail nor making Shore, nor taking prize, but endure a long painful and perilous Voyage.

For our Stone suffers itself to be overcome only by the perseverance of those Sage Caviliers of the Golden Fleece, which understand the peculiar State and general Ceremony of this Large business.

To these Sages and no other she communicates herself, yet not indifferently to all, nor always, but in a certain Season which Nature and she agrees upon things conserving in maturity the Corn being now ready to reap and the reapers solid heads coloured as the Corn and made capable of the dowry according to Geber's saying:

Philosophers that found and had this Stone,

Obtained it not till all hot blood was gone

And yet when all men thought them weak and old,

They could embrace their loves as Lovers should.

To which Age principally prudence and solidity are familiar or new which have in this time of ripeness banished all the Levity and rashness of youth, and brought all their passions to stand bare headed before them for which Cause Senior says; That a man of Spirit and long experience may easily prick his Voyage to arrive happily through this Art at the Cape of Good Hope, if he give himself wholly without discontinuance to read good Authors by whose means he shall be illuminated, and find an easy entrance to the true knowledge of this divine Secret, as is Affirmed by these Verses wherewith I will make an End.

Grey heads are they that free those Egletets

That Saturn catches in our glassy nets,

The winged feet of nimble Mercury,

Are only lim'd by grace Sobrietie.

HALLELU JAH

OF NATURE AND ART

A thankful offering of an

Enlightened Writer of the Hermetic A.B.C.

of

a well-disposed Christian Hermetic

Scholar.

Written in the Month of
November
1730

I. C. H.

OF NATURE AND ART

If we wish to know something of the inner strength
of Nature according to the measure of God, we must
look for this in every creature beholden to Nature.
Every creature must find his own particular strength
and function through the knowledge of the Universal
Substance, the highest Arcanum of the whole of
Nature and in which lies the concentrated power of
heaven and Earth.

Let us see what Nature actually is and what she is
called, and secondly how she works, or how through
this hidden knowledge the natural can be overcome
and regenerated. This knowledge is the true Alchemy
of which one reads. Now through this Art and the new
rebirth, man can contact the Q. E. of the most
secret parts of his being.

The highest and most inward strength of Man lies in
his Sulphur or Tincture, bound and held however by
the Salt of Nature. One must therefore open or
unseal this Salt of Nature, so that the tincture
SULPHUR may be freed, awaken and made spiritual,
that it may be made workable in this anxious and
careworn life. This can only come to pass as said
before through purification and rebirth, to going
back and through the 4 Elements from which it came
The new birth or tincture freed from the Salt of
Nature spiritualised and brought into its highest
tincture, SULPHUR. This separated from the false or
cursed earth, is then the highest or Quintessence
which in every body is the specificated working and
held strength. This new birth of the Universal
Substance can be achieved, for though universal it
can be individualised.

128

Art must follow in the way of Nature, for Art cannot achieve by itself, what Nature has not potentially in itself; so we see that unless we have foreknowledge, we cannot proceed to show how this rebirth can take place. If we follow false paths we cannot hope to achieve our aim and will never reach the goal. In understanding Nature lies the real comprehension of Alchemy, and we see why the wise so truly say "According to Nature so it is", and we truly work with her.

If we wish to understand the beginning of Nature and see from what God made Heaven and Earth, we must seek in Christ what Moses taught when he said in Gen. 1. "The Spirit of God brooded over the waters". We learn from this that the Prima Materia, or the first Chaotic Beginning of All Things was Water, after 6 days all creatures were brought forth and made, of which Man was the last and end of Creation," Peter. 1. 3. 5.

Mans body was created from an earthy mass or as the Wise say from Adamah, that is from a red earth, from which man received his name and nourishment, and it was called Adam by God. But this watery earthy mass or body is no ordinary earth, but a tinctured earth, full of light arid living strength, an extract of all ideas of the Spirit of the great world, after heaven and earth and all the riches of nature. That is why man is in sympathy with the world, for all the forces of nature come together in him. The Universal in the Particular, that is why the Wise say, that Man is the Centrorum and Centrum.
In Man come together the heavenly and earthly false nature. He is the Microcosmos, in him are all the

forces that go to make the Macrocosm. The Soul and Life of Man was breathed into him by the spirit of God, which enabled him to become supernatural and to make a double stand, both physically and spiritually: the body receiving strength from the Spiritus Mundi, the soul from the eternal nature grounded in God. Both are together in Man; thus is Man above all creation and is beholden to his Maker, for God wished all else to be below Him.

In this manner was man to be master over all creatures, and know nothing of death. After however man broke God's law and fell through his false earthly Will, the Light of God's Tincture left him and with it mush of his perfection and sovereignty, so that he became more subservient to the great world. God's curse not only fell upon man's self will, but on the fruits of the earth which he takes for his sustenance, and therefore partakes of inharmonious or corrupted matter. The curse of God then came over mankind and the world, and with the flooded earth came the judgment; God did not wish all to be spoilt but nevertheless disorder had come into Perfection, hence the farmer having weeds, thorns and thistles growing beside the good fruits of the earth. Hence Man had to eat his bread by the sweat of his brow, and the farmer to build and grow his crops. From this we see that still a little of the uncorrupted Light of Nature remained in the earth, for otherwise the farmer could do no good with his sowing, as also in Man remain some of the unfallen Light. Man must re-awaken this by grace in himself and nature and make a second stand, so that in this Life, if not entirely, he will in part return to the state from which he fell.

It is this Light of Grace, this Image of God in us which is the life of the soul, a beloved FIRE, of which a little yet remains in us. Though small it can become great, if we cast away doubt and believe in Christ, through which this Light of Grace can be fermented and multiplied, for through His Incarnation and Death, did He reawaken the image of God in us, so that coming nearer to Him, this Image of God can be once more attained. What remains of the light of Nature in man, is that of natures FIRE, which holds mans body together and which is ordained by God. Before the Fall this Fire in Mankind was above the elements, and therefore sufficient. This Fire could not be aroused wrongly or become too powerful, so the first man in this place knew nothing of sickness. After the Fall it came below the elements, thereby causing life and death.

As in all things and also in man, this natural fire has two qualities in itself, which are heat and cold; the heat being the fires spirit, the cold the fires body, the two qualities making the fire of nature twofold, one working towards love, the other towards anger. In love it works when heat and cold are balanced, complementary to each other, in which case it is nourishing, giving man an inborn warmth to his powers and keeping the body in good health. In anger the fire of nature is in disharmony, either cold or heat being predominant one over the other, causing strife in man's body, either consuming it with fever, or giving it a rigour caused by excessive cold. The force of nature then in man works either toward heaven in love, or towards anger in cold, even unto death. The anger of nature can also be brought about by undue enjoyment of food and drink, through a disorganised or erratic life and many other ways. When these contrasts fight, one against the other, they isolate themselves and cannot reach the harmony of God's intention.

No sickness can befall man except that which he has brought upon himself by falling away from God. For this God has created a special antidote, or specificated means to alleviate such sickness. Man has but a short span of life in which to apply and understand the specificate means of regeneration. God left a Universal Medicine through which health can be obtained, arid this is the Lapis Philosophorum, a noble Tincture, a newborn EARTH, which is but a prelude to the refined and Paradisica]. Light Wisdom of the regenerated creature, of which this Stone is a particle, for it comes from the pure elements out of which it is born.

The strife of the elements are again through rebirth, put into harmony by the One-ness of the Temperature. One can see that man with his anger and his natural FIRE, or whatever quality is predominant in his nature is torn this way and that, and to cure ill health this must be brought into harmony. In as far as man lifts the darkened Light of nature and regenerates his physical body, many great and wonderful works can be achieved by the Tincture of the Wise, called the outward Healing Balm.

As we gain the heavenly Tincture through Christ which God had originally intended for us so can we through the earthly Tincture renew the darkened light and our elementary body, until we reach our desired goal and cure our bodily ills. We cannot reach the deathlessness of Adam or the Oneness, which we had before the Fall in this world, for God willed to punish man for his sins by death. When however we have reached Gods' home for us, we must, to use an old phrase; "Whether herbs or plaster die," in order to leave our false home; then the soul wings its way again to God from whence she came. The body made from the earth, returns to earth from whence it came in which it putrifies and is made clean, until finally at the youngest last day it stands purified and becomes one with the soul for all Eternity.

As Man was created on the last day of Creation, so on the first day God created Light. Let us consider this Light.

It is nothing else but the first fiery spiritual body of the Threefold God, who in the beginning suspended a fog or mist over the waters, which with its living working strength gave it its soul and made it fruitful, God then took this living Fire, placed it in the heavens and called it Light. God then took this Light, and with its strength impregnated the earth and every creature with His Likeness, so that all created beings should have a portion of this Light.

Therefore say Salomon, "Gods' imperishable Spirit is in all". This specificated light is given to all living creatures, it is also a part of the universal and inexhaustible strength which nourishes and holds all things. This Light is the universal working Fire which the Wise call Nature or the object of all Wonder, Spirit, Sperma, Hyle Archaeum, or the Universal Spirit, for all the strength of nature is in it, and we see in this spiritual Wisdom the bodily earth. The Hermetists say that "as above so below". As long as this spirit hovers in its own sphere, it is universal and can draw to itself diversity of forms in three different kingdoms, Metallick, Vegetable and Animal, but as soon as it has specificated itself it is no longer universal and becomes one with the body which it inhabits. All creatures have their beginning in this spiritual body and through it comes the magnetism of nature, for the creature has its root in spirit from which it broke off and for which it continually yearns; and is drawn too and maintained and also made essential. It augments and multiplies vegetation.

This Light or Universal Heavenly Fire of Nature, the life and movement of all things remains as long as God wills it, nourishing vegetation; for it is an imperishable strength which the curse of God did not lessen over the Earth and its substance. For were this not so the heavenly fire could not have increased the cursed earth nor given fruit to vegetation, and the world would have come to an end. But it gives continuous motion and life and its souls' strength until God decrees the end of the world and brings it to Judgement.

As this Light was first from God in heaven, the planets obtained this Light by proximity, which gave them life and passed on this life to their lower physical bodies. The life of the Planets is dependant upon continuous movement, ensured through the influence of the Heavenly Light. Therefore Bechard says; "The Sun is either hot or cold to itself, but its movement is naturally hot."
In their circular movement or path they form a certain figure and stand in the heavens in the following order: SATURN, JUPITER, MARS, SOL, VENUS,

MERCURY, & LUNA; and while the Planets influence
that which is below not above, it is apparent that
all their strengths are given to all births; the
source of the Seven spirits, named the Seven Forms.
The whole birth of nature comes from the Seven
Forms, and through them take their characteristics,
although in each thing only one Form is predominant,
as Chortolassaus writes. As SATURN is the highest
planet so he is the first to catch the influence of
the spiritual light; which he attracts through his
own cold property and which he then shares with the
other Planets, of which the Moon is the last of the
assembled planets to receive the influence of the
spiritual light, the latter sharing it with the
elementary world to which it is closest, the Earth.
The lower element, our world, has two bodily
elements water and earth, as against the above two
spiritual or working elements, fire and air, which
when broken off their spiritual home become a
substance or body, for the fire rests in the air,
and the air in the Water.

They are both in the earth which is the centre of
Nature and which fully encloses all the elemental
powers, for they are mixed in an earth and rest in
it.

Before the Fall the elements stood in equal harmony,
for while the powers were in harmony there was no
strife, therefore the Paradisical fruits were
without blemish. After the Fall however, the
elements were put off their proper temperature, and
thus came to strife, through which the fiery Light
spirit took substance and thus fell into matter, and
in so doing enkindled the fire in itself, became
partly false, and because of that put the air out of

its right proportion, and its weakened fire became mixed with a distasteful dead water, which made strife inevitable in the elements. No species in its own kingdom can reach the highest point of fulfillment, for the spirit of Nature through the false SULPHUR and the powerless water cannot perfect nor carry it through to a greater perfection.

Although the 4 elements strove fiercely against each other after the Fall, they had to harmonize and come together in the birth of all things as One Whole, which could not happen unless they came from one Materia, which separated from the watery chaos. There is a close connection between the elements, for one cannot function without the other, but indeed energizes the other. In the ground of Nature they are ONE.

The earth is a coagulated water, and the water a dissolved earth, but without air there would be no movement or life whether in water or in earth. For in fire alone is the movement and life of all things. The Fire is nothing else but a thinly made water, and the water a contracted or drawn together Air. Now Earth, Water and Air would be dead and non-active without the Fire which has two extremes, namely heat and cold. With its heat it dissolves all, and makes the earth into water, and the water to air. With its cold it congulates everything and turns the air into water, and the water to earth. So much we know, that the fire with the heat of its agent, is the first cause of the Holy Spirit, contrary to the action of the cold, for this fire with its two different properties, through its influence in the left over Planets, brought forth all things in the World.

In this knowledge lies the greatest mystery and Source of our whole philosophy. It is however the fire of nature on the 4th. day of Creation that was made into a body, and according to its property (two-fold) was placed in the firmament of heaven, and encounters the two great Lights, namely Sun and Moon, which send their strength and pass their influence on to all the other planets and stars and the creatures in the three kingdoms. The Sun consists of the warm central fire and from the Moon the same central cold fire, one being dependent on the other.

Without the heat of the sun the cold central fire of the moon would be enclosed in death, while the warm central Sun is the movement of the cold life, but the warm central heat of' the Sun would not be fruitful of itself, for its heat would burn and destroy without the cooling temperature of the cold central fire for nourishment, but both central fires would be dead were it not for the universal Spirit of the world which is in between both, energizing and bringing things to life.

These two central fires with their respective properties and Holy Spirit constitutes the three spiritual or holy principles in the upper elementary world, as in the lower elementary world all birth comes into a corporeal body from the centre. They are the same essential parts of all forms, namely SALT, SULPHUR and MERCURY, without which the Universe cannot give birth or work.

In the lower elementary world these 4 elements in them; they are in fact a seed or birth of them; for out of 4 elements come 3, which happens through a separation of the weakened water. Nature in the beginning had need of the 3 principles which through mixing and uniting, was necessary to an outward birth. After the conception and union of the same, there comes to birth a radical being who throws away any superflouosity for the imperfection of all things is caused by not having been sufficiently processed. The 3 principles differ in the matter of working but in nature they are in reality but One, can be mixed one in another as we see in the birth of things that eventually go back to one principle. In the lower world all creatures are constituted from the cold central fire, of which the salt is the dry basis. This is under the influence of the first three fiery forms enclosed in the earth. Its working is astringent therefore sulphur and mercury is active, and draws together, coagulates these forces which then become corporeal. The Sulphur or the natural inherent warmth is the second principle and is the warm central fire, an oily substance or essence; through this essence it holds the salt together in which it overcomes a body; in it the sulphur is a subtle earth which can easily become a hard and fiery earth, wherefore it is twofold, both heavenly and false.

Through the cold central fire or salt it becomes matter, secret and closed and changed to a fiery earth. When however it becomes revealed and awakened through the third Principle the Holy Spirit gives forth a spiritual power and a working life which brings all things to proper growth and fulfillment. Finally the third principle is MERCURY or the first beginning of moisture the beginning of all growth. This, of the three principles is the most fleeting

heavenly essence, for it comprises within itself all
the powers of the upper elementary world, which
comes down to the lower world, so that it Is called
the Wisdom of God's abode, without Him is the salt
and sulphur dead and without energy; when however He
works in them and opens them, they ferment into a
living life. Through this comes a working and
multiplying strength. In this strength of power the
mercury is a purer spirit in which form it cannot
remain in the lower world, unless it is caught
through the earth, in which a bodily and fleet-ing
salt is mixed, so that it can through its nature
reach from spirit to matter, and can come to all
creatures.

When it receives help from its persevering fire or
sulphur, it should not be burnt out and made into
air, but increase and strengthen its life. When the
above two principles, the 2 central fires tolerate
each other, the mercury rests in them spiritually,
for with its warmth, it opens them out, and with its
cold, closes them, for it stays with all things in
an active or spiritual manner, or in a passive
moderate manner; the passive part into the
multiplication of a body, in which the spiritual
part rests.

From this we see that not one principle can give
birth to anything, but that in every birth the three
principles must come together. The salt gives the
body, the sulphur the life, the mercury however
cleans both and gives it strength to live, if it has
awakened the central fires and inwardly truly come
together, so that from the three principles the 2
working central fires are born. For through the
mercury the opened salt presents the working cold

central fire, and in as far as birth is concerned, the increasing or passive power lies in women, the mercury resolves, and the sulphur is fugitive and the warm central fire's property, and the male property activates the cold central fire, then it comes to full birth.

When these central fires have taken hold of each other, in their most inward essence, the goal is reached and brings about a birth in the world which means that from the inner ground, each creature has obtained its own understanding, in the tenth number, for every body takes its natural source from the Li. elements; from these are born the 3 principles, from these flow the 2 central fires, each after its own fashion, working towards one simple number, and with the same making every birth possible, for nature when it has obtained its completeness can work no further, but rests still and quiet. It is then called the Tenth number of' the Wise, the complete Number. God himself fulfilled the tenth number and gave it to all His creatures, then rested. So this number is called Holy, for in it God Himself rests. When philosophers speak of this number, they understand the Light, which is the first root of all things. When they speak of the second, they mean the 2 working central fires, namely heat and cold, salt and sulphur, then they mention the third, they indicate three principles, Salt, Sulphur, and the moving Mercury, and the fourth the Li. elements of which fire and air are spirit, and two water and earth are corporeal. As these outward births of nature come in a spiritual manner from ONE out of Light, so also their strength goes again into the ONE after they were enclosed in a physical body. As the tenth number is the highest in its inward essence, so is the false and outward principle the number 7, for everything that God created, which can be found in the world, can be found under this number. In heaven we have seven planets, which Light God has Himself created, these through depending on the First Light, their essence taken from it, are dependant on one another. Their Essence being obtained in this way, is nevertheless different, in as far as their greatest strength lies in the two planets, the Sun and the Moon, and in their two ways divide themselves into specificated Substance. The Sun being the warm central fire of nature and the Moon the cold central Universal Fire.

Through the highest strength of Light, from which they constantly spring, they give in turn light and splendour to certain planets, and to others their exalted strength in the lower world, and to the three universal kingdoms, their activity, to the accomplishment of production, conservation and multiplication; and while every creature has its birth under one of the 7 planets of God, each planet takes his own circular strength to himself, which may be either form the Suns or Moons property, for these 2 principle planets have the universal powers of the whole of nature, in which the other planets are particularised, of which the country-man speaks in his Cabal Chemica.

Go we further into the most hidden part of the earth, as in the mineral kingdom, we find that it harmonises with the upper in the 7th. number. For as seven planets reign and govern in heaven, so they govern the seven metals in the world, which is under its dominating planet, and enveloped in its own spiritual power. What is therefore the sun in heaven is gold in the earth, as it has enclosed itself in its spiritual strength. This applies to the other planets.

We find that the vegetable kingdom is also in sympathy with the seven planets, no tree or vegetable or anything else in this kingdom can be found which is not under one of these heavenly influences. A certain planet predominates over one thing, not as strongly as in the mineral kingdom, for in the former the power is more strongly concentrated. We come at last to the animal kingdom, which harmonises with the planetary kingdom in the seventh number. As man however the Centrum Centro the Q. E. of the great elementary world after heaven, earth and all kingdoms were created by God, so does he contain the lower and the upper astral powers fully in himself. Therefore Sun and Moon and all planets are in him, for he is the Microcosm in which they work in his principal organs. We see the most harmony and working of the Sun in the heart, the Moon in the brain, Mars in the gall, Venus in the kidneys, Saturn in the liver, Mercury in the lungs, and Jupiter in the spleen, to which they are all attached and drawn by the Light.

The Light is the heavenly Universal Fire of Nature and the first working Being of whom Morienus said "Of whom there is no beginning save his Creator." Without the penetration of the Light the Regnum Astral cannot of itself exist, in so far however as it favours the Influence, it not only obtains strength through its own existence, but increases and works through to the lower world to the three kingdoms, with sufficient Light and Life force to uphold them. Upheld by the Creator, specificated or brought out from the Universality, and after the working manner of each subject with which it has joined, it is mixed in Nature and existence.

When the Light Power is joined with the Regno Animali, it takes on the animal nature, and in like manner hardens to flesh and bone. The same happens with the Regno Vegetablili, its property being mixed in fruit, wood and foliage, the same with the Regno Mirierali, which takes to itself the mineral nature and existance in stone, minerals and Metals.
As all creatures have taken their source from the Light, so naturally in the same way they should seek their life force and strength through it also. All Creatures draw their magnetic powers from the upper Forces, through the salt which is all found partly in spirit partly in body; all the more when a creature participates in the fugitive salt which enables it to stand and to draw a greater strength from the upper astral spirit, which looks upward to its own fire, and which from the fiery salt makes all creatures grow in their own fashion, for in the salt lies the magnetism of the heavenly powers, which shows that God has endowed nature with perpetual growth and existence.

The salt of nature is in all creatures, contrary to Spirit which cannot unite with a central body, and which cannot remain in it. We must further consider how the Light can rise from its spiritual nature and unite itself to creatures and how it observes the grade of fixation. The elements in their contrarieties as fire and water are fugitive and concentrated, and can be united in earth. They serve the elements that can make extremes meet, for by transmutation one can have success with the other.

When nature wishes to change fire into water she works upon air, in which she draws together fire and air and has commerce in water. Therefore will she make air into earth, happening as it does through water, for the air next in corporeality, is the water through which condensation (thickening) is brought into earth. If nature wishes to turn fire into earth, she uses two elements, being able to use extremes of function through their contrariety; so firstly she brings the fire to air through an extension, this again draws them together and mixes them in water, which becoming more dense goes into earth. In this wise do the elements of nature mix agreeably together, for through the middle element, they can once again become united to the ONE. Here we have the Microcosmical prelude of the new Heaven and the new Earth of which much is written.

Even after the Universal Fire has divided itself into two properties it remains spiritual. But a Spirit which cannot unite itself with a body and stay with it, would not benefit the lower world, unless it gave itself further to the elements and to a middle nature, between Body and Spirit, so that it could incorporate itself in creatures and remain with them. It is however next in the fire, and then Air, so that it can circulate freely in the lower world, and can be seized in the Air element, in which it still rests as spirit, while the courser fire, takes a more subtile and spiritual salt body to itself which the philosophers call Saltrum Universali, in which living creatures stand, and of which Sendivogious says "In the Air is a secret life food, which at night we call DEW and in the day WATER am raresactam." For through the suns warmth, the air is drawn up and thinned, contrary however to night when through the coldness of the moon, it is drawn together, thickened and changed into moisture and dew.

Whether creatures draw this air for their conservation one cannot be sure, though they enjoy the animal kingdom particularly; Man his Adam-hood, for through his own fire and warmth, and through his nature and being, he digests and transmutes it, so also the outbirth of the vegetable and mineral kingdoms, which is stimulated in water by the false light. They are one more grade nearer to solidarity. The grade nearer to matter than air is water, in which is found the Holy Spirit, or a subtle salt-nitre, which is held (constrained) in a seed and makes it swell, otherwise it would burst or easily change into water.

In the form or body of water are certain creatures composed of Vegetable and mineral kingdoms; the upper life force with its growing power has through the water a universal sal nitre, indeed lives in it, for it is through this nitre that it can grow and remain. It is in this heavenly or universal salt that this acid property has two central spiritual and specificated (determined) centres. From this fugitiveness and acid salt, comes the mildness and sweetness of alkali in so far as changes come through the mixing of the volatile with the fixed or Acidum with the alkali through the water which is the instrument of the mixing, cleans itself and in every body produces the strength of the Seed. This happens through the sal nitrum coeleste in Primem Materiam spermaticum for when they ferment, the volatile is united to the fixed, until at last both give themselves completely to fermentation, in which the heavenly sal nitre and the bodies alkali salt is bound together, and throws away any superfluous water. With the salt at last being earthly and terrestrial and so constituted that it is in every creature, and also brings the elements into One, or as Hermes says, "Strength can be held in so far as it is transmuted in earth." In such a manner rises the heavenly Light Spirit, or the Universal Fire of Nature, in the elements for the sustenance and multiplication of life. As everything by virtue of life has warmth in it, so everything pertaining to death has cold, which we see in summer and winter.

In winter when it is damp the cold central fire predominates and we take it for truth that his cold and astringent quality of the earth closes the life and growth of it, it partly holds it back, partly suffocates and kills; so it affects other creatures who have not sufficient warmth or fire in themselves, to combat this cold and deadly property, but in summer when the central warm fire triumphs, we see how the warmth penetrates the earth and opens it up, it awakens the fire in creatures and so once again starts their growth, for the light having two extremes, heat and cold, life and death, these by their own natural inner fires warmth have a long life to look forward too, while those who cannot combat the cold through a deficiency of warmth, soon come to death and corruption, or through sulphur have a cold salty body. Anyone having a predominance of either heat or cold, death is the result, a healthy body must have both qualities in harmony.

Finis.

LIBER TRIUM VERBORUM OF KING CALID

B. M. Sloane **3506**

Of the Quality of ye Philosophical Stone.

The Stone out of which this work is made has in itself all the Colours, for he is White, Red and more Red, Yellow and most Yellow, of a Celestial colour Green and heavy.

In this Stone are the 4 elements for he is watery, airy and fiery and terrestrial. In this Stone the calidity and siccity is in occulto and the humidity and frigidity in manifesto, therefore we must hide the manifest; that is we must make manifest that what is occult, for that what is occult, namely calidity and siccity is oil, and this oil is dry and this siccity Tinges and nothing else, for alcali tinges and nothing else. That which is in manifesto, fridgid and humid is a corrupting aqueous fume therefore it is fit that the frigidity and humidity be equal with calidity and siccity, also that they fly not from the fire for betwixt frigidity and calidity is one particule which is hot and dry therefore the frigidity and humidity must receive the calidity and siccity which was in occulto and be one substance for that humidity and frigidity is a corrupting substance of which it is said that the aquose and adustibe humidity corrupts the work and tinges it into blackness, and this infirmity must be destroyed by fire and by its gradus.

Of the Property of the Stone.

This is the book of the 3 Words, the book of the
precious Stone who is an airy volatile frigid and
humid aqueous and adustive body, and in it is
calidity, siccity, frigidity and humidity another
virtue is in occulto and the other in manifesto.
Also that which is in occulto be made manifesto and
that which is manifest may be made occult by the
virtue of God and by calidity, for the Persian
Philosopher says that frigidity and aqueous and
adustive humidity is not amicable to calidity and
siccity for calidity and siccity destroys the humid
and adustive aquosity by divine virtue and then the
Spirit is transmuted into a noble body not flying in
the fire but like an oil which is a living
multiplicative Tincture everlasting and a precious
Sol. of the Occult Calidity and Siccity exerting in
Humidity and Frigidity.

The wonderful work of the 3 Words is the work of the
precious Stone in which the aquose and adustive
humidity and frigidity and in the same the occult
calidity and also that what is read of the three
Words is by some otherwise understood that all
people might not understand the cause in the 3 Words
this is sought in humidity and frigidity in which is
the occult calidity and siccity, and that we must
know that we may make of the manifest an occult, and
an occult of the manifest, and the occult is of the
nature of Sol and fire, and it is the most precious
oil of all occults, and a living Tincture and a
permanent Water which lives always, the Vinegar of
the Philosophers, and a penetrating spirit, and it

is a Tinging and revivefying occult, which rectifies and illuminates all dead things, and makes them rise, and then its calidity and siccity does not fly from the fire but the aquose and adustive frigidity flys from the fire and destroys itself.

Of the Conversion of the Spirit into a Body and the Body into a Spirit.

That we may make all manifest namely the occult into a body and the body into a spirit then a friendship is made betwixt the frigidity and humidity, calidity and siccity. Therefore the Persian Philosophers say that it is a wonder how it should be but by the Power of God it can be with a soft temperament and moderate gradus of fire in the space of 2 and 7 days, for of 3 two are understood, and of 2, 5; but 3 is not understood and these are the 3 Words precious occult, and apart, given not to ungodly infidels but to the poor, from the first to the last man.

Of the Planets and their Images, and of the Operations existing in Mercury.

I say that in Mercury are the works of the Planets and their Images in their own places and the work in their own times for in the FIRST MONTH in the womb when the sperm is received by the matrix, then Saturn operates, congealing by its frigidity and siccity, the matter into one mass.

In the SECOND MONTH Jupiter operates digesting by its calidity into a fleshy mass which is called Embrio.

In the THIRD MONTH the mass operates and by its calidity and siccity divides, sequestrates the mass and divides the members.

The FOURTH MONTH Sol like a great Lord immettes the spirit and gives life.

In the FIFTH MONTH Mercury operates who makes the holes and spiracles.

In the SIXTH MONTH Venus disposes and ordains the eyebrows eyes and such like.

In the SEVENTH MONTH Lunar by its frigidity and humidity operates to bring forth the Foetus and if it should be born then it is debilitated.

In the EIGHTH MONTH Saturn operates again, by its frigidity and siccity constraining or constricting the foetus and if it is born then it could not live.

In the NINTH MONTH Jupiter works again and by its calidity arid humidity nourishes the foetus arid when the 9th. month is complete then the foetus is born and lives, and there are three Words, the Water preserves the foetus for three months, as also for three months who makes also the blood in the navel and condenses the same after the birth into milk, for the infant cannot be born before the aireal flatus are gone.

Of the Observation of the Planets in the Work of Alchemy.

From this 3 months you must with acute ingenuity compose and extract two for two are not three understood, therefore all who intend to understand this Art, must sharpen their enginuity to open the Treasury of these 3 Words in which is hidden the whole operation and power of the Stone, in which is the Calidity and Siccity, which siccity in a living Oil and a living Tincture and is a tinged siccity, and a profundity of tinctures and this is the conjunctive calidity and humidity, and all from the Beginning seeing this Word, did not know it, and they who heard of the 3 Words did much wonder and the position is this:

From the beginning of conception till the nativity of the infant, every planet in his place shows an image by the divine power, Creating it also. And I Rackadebi say, and it is true, that in all chemical works every planet in his place shows an image till the compleatment of the operation, and then Alchemy is born artificially, but this is truly generated naturally according to the planets, like God did show to the first man, having naturally the nature of all Tinctures, and also Mercury is born having in him the 4 elements and the nature of all Tinctures, according to his gradus and in this work of Alchemy many err and few come to an end for in this work is the DANCE OF THE MOON AND THE CIRCLE OF THE SUN TO THE 3 GRADES, the first weak, the second strong, and the third perfect, and THE THREE TERMS, the first when Sol enters into Aires and is in its exaltation, secondly when Sol is in Leo, the third when Sol is in Sagittarius; but the circle of the Sun is of 28 years, in 19 years in his mines arid other tables of Alchemy Chimia is compleated; for by the number of the dances of the Moon we find the grades and from one in two grades, CLXIII till XXIV and we find in the circle of Sol 7, Understand for by this gradus the work of Alchemy is compleated.

The Exposition of the Three Words.

Let us come again to the Exposition of the 3 Words in which the whole Art of Alchemy consists, it is said that the water preserves the foetus in the matrix for 3 months, the air for 3 months, and also the fire for three months, and this is said for the Mercury by similitude, and this obscure word and term is opened to understand the truth, for there is another nature in a child bearing woman and another in Mercury, but by similitude of the heat which is found in the matrix the fire is attained (estimated) who is of 32 gradus. Therefore that third word is obscure of which is said that the fire PRESERVES, and many feel in this, for of the 3 take 2 gradus, and out of this 2 gradus, the other are extracted in 32 patiently, and in this gradus is all the Third Word explained of which is said that the first gradus compleates the Water, and Air the second gradus, compleates all that we have said and this is the gift of God.

Of the Gradus of Fire.

The Philosopher of the King of Persia and the Roman Prince says:

Also divide the 3 Words in 2 parts and this 2 parts divide again in 2 parts. And over this 2 divide 32 grades, which are the terms of fire, and are called the particles of fire, this is found in the portion of the work which is divided into 32 parts, and are called Almes (?). All this gradus are spread over

the 2 first parts which are 2 terms and the 32
gradus are packed into 14. parts the first gradus is
the particle of fire one albechir, and is (one) and
only simple and is most none, and it is a gentle
fire, and with that fire we begin to comprehend the
Mercury to the Red and also 2 words are compleated
in 6 maenchen, after this the 3rd. word is explained
which is obscure and in which many feel and lose
their senses, the Persian Philosopher says:

Let us divide this in the middle, the mediate is of
3 maenchen and this mediaty is governed by 2 gradus
which are two particles of fire and also are
compleated this work in 22 maenchen and this is the
first term without any error, the second term 16
maenchen and is governed by 8 gradus of fire, and
the third term is of 20 maenchen and is governed by
16 gradus that is particles of fire. the fourth term
is of 24 niaenchen, and is governed by 32 gradus go
fire, Adranus and all the Persian Philosophers say
by God and his Holy Name blessed, for this is said
of the temperate fire over the 3 words, of the
nature of a child-bearing woman, to the comparison
of the fire of Mercury. All these two terms are
divided in the middle for they are both 32 maenchen
and are 7 dierchen and in the end of the first term
open the treasure and project what you find, which
if it dances and smokes over a hot plate, then it is
not enough, therefore bring it to the fire of 16,
which has in it 8 gradus of fire, open again the
treasure and put it over a red hot plate, and if it
dances and smokes it is not enough, therefore bring
it to the fire of 20 which has in it 16 gradus, open
again the treasure and if it smokes still it is not
yet boiled, bring it therefore to the fire of 24
maenchen and 4 dierchen which has in it 82 gradus of

fire and now you will have a precious fusible Stone, golden and red. In this hour let God be blessed, and his Holy Name which is blessed above all names, because of this Holy Gift.

Finis.

THE PHILOSOPHICAL CANNONS OF

PARACELSUS

B. M. Sloane **3506**

1. That which is near to perfection is easily brought to perfection.

2. The imperfects are by no means brought to perfection before they are deprived of their feculent Sulphur and Terrestrial thickness which is mixed to the Mercury and Sulphur, this is a perfect Medicine.

3. To make fixed the imperfect without the Spirit and Sulphur of the perfects is impossible.

4. Heaven of the Philosophers resolves all things in the first matter that is Mercury.

5. He who intends to reduce metals into Mercury without philosophical Heaven, or the metallic aque vitae, is cheated, for the impurity of Mercury may be seen in all other dissolutions.

6. Nothing is fixed perfectly which is not mixed indissolubly with the fixed.

7. The fusible gold may be altered and turned into blood.

8. For the fixing of Silver, it must not be turned into powder or dissolved into water, for this is

destruction, but it must necessarily be reduced into Mercury.

9. Silver may not be turned into Gold but by the Philosophical Stone, except it be reduced into Mercury. Also is done with other Metals.

10. Imperfect bodys are brought to perfection and into perfect Gold, when they are first reduced into Mercury, adding to it white or red Sulphur.

11. All imperfect bodys are brought to perfection by reducing them into Mercury and afterwards by boiling them with Sulphur and appropriate fire, for then they are brought to Silver and Gold, and they are cheated and work in vain who intend to make Silver and Gold otherwise.

12. The Sulphur of Mars is the best, and this joined with the Sulphur of Gold makes a Medicine.

13. There is no Gold generated except it be Silver before.

14. Nature makes and generates minerals by degrees, also out of one root are generated all metals till the end of all which is Gold.

15. Mercury corrupts Gold and resolves it into Mercury and makes it volatile.

16. The Stone is composed of Sulphur and Mercury.

17. If the preparation of Mercury is not taught by an expert Artist, it win never be found out by the reading of books.

18. The preparation of the Mercury for the Philosophical Menstruum is called Mortificatio.

19. The Praxis of this Arcanum goes beyond all secrets of Nature and it be not revealed or taught, it will not be learned out of books.

20. Sulphur and Mercury are the matters of the Stone; therefore the knowledge of the Mercury is necessary for the election of a Mercury fit for the Work.

21. There is hidden a Mercury in a body prepared without any other preparation, but the Art of extracting is difficult.

22. The Mercury may be fixed and turned into gold and silver for the compendium or abbreviation of the work.

23. Fixing and congealing is one work, of one thing, in one vessel.

24. That which fixes and congeales the Mercury tinges it also in one and the same practise.

25. Your grading of fire are to be observed in the work, in THE FIRST the Mercury dissolves his body, in THE SECOND the Sulphur drys up the Mercury, in the THIRD AND FOURTH the Mercury is fixed.

26. Things radically mixed, afterwards grow inseparable; like snow mixed with water.

27. Divers simples, put into putrefaction produce divers others.

28. The form and the matter must necessarily be of the same species.

29. The homogeneal Sulphur is of the same nature of which is Sol and Luna, and this Sulphur produces pure gold and Silver not in that form as it is seen with eyes, but as it is dissolved in Mercury.

30. Without the philosophical dissolution of gold into Mercury, may be extracted out of gold a fix sort of unctuosity, which takes the place of a ferment generating Sol and Luna and what is done by a way of abbreviating which Geber calls Rebis.

31. Metals resolved into Mercury are reduced into a body again by adding a little quantity of ferment for else it retains always the form of Mercury.

32. The leaven of the Tartarus of the Philosophers which reduces all metals into Mercury is the metallick aqua vitae of the Philosophers, which also they call dissolved faeces.

33. Sulphur and Mercury are of the same homogeneal nature.

34. The Stone of the Philosophers is nothing but Gold and Silver exalted into a higher Tincture.

35. Sol arid Luna by themselves in their own species have riches enough. Them you must reduce into the nature of a ferment. This mass may be multiplied.

36. The most extremeries in Mercury are two, namely crudity and most exquisite decoction.

37. The Philosophers observe that all dry things quickly imbibe their humiditys.

38. The altered calx of Luna quickly imbibes his Mercury, the fundament of philosophic minerals.

39. The Sulphur is the Soul, but the Mercury is the Matter.

40. Mercury is congealed into an imperfect body and goes in the same species of the imperfect body by whose Sulphur it is congealed.

41. To make Sol and Luna with the Sulphurs of imperfect bodys is impossible for everything can give no more but what it has.

42. The Mercury of Metals, is the feminine seed, for by projection it goes through the qualities of all metals till gold.

43. For the extracting of the red Tincture, the Mercury must be animated with the ferment of Gold, and for the White with the ferment of Silver.

44. The Philosophers work is quickly done without any expences and that in every place, at all times, if they have but the true matter.

45. The Sulphurs of Sol and Luria fix the spirits of their species.

46. The Sulphurs of Sol and Luna are the true masculine seeds of the Stone.

47. All which have power of fixing must be necessarily permanent and fixed.

48. The Tincture giving perfection to imperfects is made out of the Fountain of Gold and Silver.

49. They who take the Sulphur of Venus are cheated.

50. Venus has naturally nothing which is useful or which can serve in the great Spagyrical Work.

51. Sol converted into Mercury before the conjunction with the Menstruum cannot be a ferment, a soul or a Sulphur.

52. The Work brought to an end by reiteration is made fiery.

53. In the abbreviation of the Work the perfect bodys must be reduced into a current Mercury which can rightly take the ferment.

54. The preparation of Mercury by sublimation is better than that which is done by amalgamation, but note that you must revive it.

55. The Soul cannot impress a form but by the help of a Spirit, which is nothing but Gold turned into Mercury.

56. The Mercury receives the form of Gold by the mediation of the Spirit.

57. Gold resolved into Mercury is Spirit and Soul.

58. The Sulphur of the Philosophers, Tincture and Ferment all signify one Thing.

59. Vulgar Mercury is made equal to the nature of the Mercury of bodys.

60. The ferment makes the Mercury ponderous.

61. When the Mercury Vulgar is not animated or without a Soul it is then not fit either for an universal or particular operation.

62. Now the Soul is impressed into the mortified Mercury.

63. Sol may be prepared into a ferment also that one part of it animates ten parts of Mercury, and this work has no end.

64. The Mercury of the imperfect bodys takes place of that vulgar Mercury, but the Art of extracting it is difficult.

65. The vulgar Mercury is turned into Gold by projection of the Philosophic Stone, therefore it may be exalted and made equal to all Mercurys of bodys.

66. Vulgar Mercury animated is a great secret.

67. All Mercury of metals by abbreviation of the work are turned into Gold or Silver.

68. Humid and gentle heat is called the fire of Egypt.

69. Note. Luna is not the Mother of vulgar Silver, but a Mercury endued with some quality of a Coelestial Luna.

70. The metallick Luna is of a metalline nature.

71. Vulgar Mercury takes on feminine nature because of its sterility.

72. The Mercury of the half minerals show the nature of Silver by similitude.

73. All things are produced out of Sol and Luna.

74. Man and Woman, that is Sol and Mercury congeal together.

75. Vulgar Mercury without preparation is remote from the Work.

76. Four parts of Mercury and one of Gold which is in the place of ferment make a matrimony.

77. The solution is done when Gold is resolved into Mercury.

78. Without putrefaction there is no dissolution.

79. Putrefaction lasts till Whiteness appears.

80. It is a great secret to mundify the Mercury with which is prepared the Menstruum in which Gold is dissolved.

81. Mercury resolves the Gold in form of a water, that is into current Mercury like it is itself.

82. The dissolution is the beginning of congelation.

83. Sol dissolved into running Mercury in a short time remains in that form.

84. The ferment drys up the Mercury and makes it ponderous and fixes it.

85. The Sol of the Philosophers is called a Fountain.

86. The Matter by power of putrefaction is converted into a part which is the principle of congelation.

87. There is a compendious way by which the Sulphur of Sol and Luna is extracted by which all Mercury is fixed into Gold and Silver.

88. The Matter must never be removed from the fire that it may not grow cold or else it is spoiled.

89. When the Matter comes to be black then give the Second grade of fire.

90. The Washing of the Philosophers is but a similitude for the fire only perfects all.

91. Poison and stinkingness is taken away only by the fire for it is that which absolves all.

92. Fire by its penetrating arid acute virtue cleanses more than any other water.

93. When in any vegetable thing the heat or colour is extinguished there follows death.

94 & 95. The Spirit is the calor (colour?).

96. When the Matter is brought to Whiteness then may it not be destroyed.

97. All corruption of things is noted by a mortal poison.

98. The Glass or vessel is called Mother.

99. The virtue of the Sulphur may 'be extended to a certain term.

100. You must observe the question why the Philosophers call their Matter a Menstruum.

101. Sulphur dissolves the name of a form but the Menstruum the name of the Matter.

102. The Menstruum represents the little and inferiour Elements, namely the Earth and Water, Sulphur the Superiour as Fire and Air are the agents.

103. When you break the shell of the egg also that the chick comes out then it is killed, also if you open the vessel also that the matter feels the Air, then it is all spoiled.

104. Calcination done with Mercury in a reverberatory is good.

105. The methods of the philosophical stile must diligently be noted, for by sublimation they understand the dissolution of bodies into Mercury by the first grade of fire, which is followed by the

second operation which is the inspissation of Mercury with Sulphur. The third is the fixation of Mercury in a perfect and absolute body.

106. There is an infinite number of Errants who do not allow of Mercury as it is in its form mixed by the calx of perfect bodys, to be the matter of the Stone.

107. The White Medicine is brought to perfection in the third degree of fire, and this degree you must not transgress in the making of the White Medicine or else you will destroy the White Work.

108. The fourth degree of fire makes the matter red, and there appear divers colours.

109. The Work after White not brought to a high redness is imperfect not only the White but also the Red Tincture.

110. After the first degree of the Persian fire the matter becomes more powerful.

111. The work is not brought to perfection except it be incerated arid made fusible like wax.

112. The work of ceration is done by addition of 2 and 3 parts of Mercury which gives the being to the Stone.

113. The inceration of the White Medicine is done by the White Water or the Mercury animated with Luna, but the inceration of the Red Medicine is done with Mercury animated with gold.

114. It is enough when the Matter after inceration remains like a paste.

115. Reiterate the inceration till it has the right consistence.

116. When the Mercury with which the inceration is done, flys away it signifies nothing.

117. The Medicine right and duly incerated explains enigma of the King coming out of the Fountain.

118. Sol. reduced into its water or first matter, by means of the vulgar Mercury, if it grows cold it is spoiled.

119. The Philosopher takes the matter prepared by nature and reduces it into the first matter, for everything is reduced into that out of which it hath its original, like Snow is mixed with Water.

120. The Wise men bring years into months and months into weeks and weeks into days.

121. The first decoction of Mercury done by nature is the only cause of its simple perfection beyond which it cannot come, but you must help this simplicity; seminating gold in its own earth which is nothing else but pure Mercury, which is by nature a little and not perfectly digested.

122. In the second decoction of mercury the virtue of mercury is ten times augmented.

123. The Stone of Mercury is made by reiterating the decoction adding to it gold and also man and woman are twice boiled.

124. Sol must be added to Mercury that it may be turned into Sulphur and then it is boiled into the Physical Stone.

125. Also some contemplate the philosophical Mercury, yet do they not know it.

126. Every Mercury of what original so ever represents the matter of the Stone, taken in a due manner.

127. Everything is the subject of the Stone out of which Mercury may be extracted.

128. All who understand the writings of the Philosophers according to the letter are cheated, for they affirm but one Mercury.

129. (is missing in the original M.S.).

130. One Mercury exceeds another in more calidity, siccity, decoction, purity and perfection who without corruption of the form must be prepared and purged of his superfluities in which consists the secret of the Stone.

131. If the preparation of vulgar Mercury were known to Students of this Art there would be no other Mercury sought for, nor any other aqua-vitae nor any other Mercurial Water for the common Mercury contains all this.

132. Every metallic Mercury by successive degrees may be brought and exalted into the quality of any Mercury of bodys.

133. The vulgar Mercury 'before due digestion is not the philosophical Mercury but after preparation it is called by that name containing in him a true way and method of extracting the Mercury out of other metals, and it is the beginning of the Work.

134. Prepared vulgar Mercury is the metallic aqua vitae.

135. The passive Mercury and the Menstruum do by no means loose the external form of Mercury.

136. Whoever uses in place of current Mercury any sublimate or calcined powder or precipitate is cheated.

137. Whoever resolves Mercury into clear Water for the making of the great Work are in error.

138. To make Mercury out of limpid water is in no bodys power, but only in the power of nature.

139. Necessarily in the philosophical work, it is that Mercury crude does dissolve gold into Mercury.

140. When Mercury is reduced into water it dissolves gold into water and in the work of the Stone it is necessary that it is dissolved into Mercury.

141. The Sperm and the Menstruum must be alike in the external form.

142. It is said in the doctrine of the Philosophers that necessarily he must moisten the nature, but if the menstruum is dry, there is no dissolution hoped.

143. You must take the Seed of the Stone in the like and near nature of metals.

144. It is highly necessary that the Seed of the Philosophical Medicine is like vulgar Mercury.

145. The highest secret of all is to know that Mercury is both Matter and Menstruum, and that the Mercury of perfect bodys is the form.

146. Mercury by itself does, nothing in generation.

147. Mercury is the elemental earth in which gold is seminated.

148. The Seed of Gold is indued with multiplying virtue.

149. Perfect Mercury seeks for the work of generation a woman.

150. Every Mercury consists out of 2 elements, the Crude out of water and earth, boiled out of fire and air.

151. If anyone will turn Mercury into a metal, then you must add to it a little ferment that it may be to such a degree of perfection brought as you please.

152. The greatest Arcanum of the work, is the physical dissolution into Mercury, and reduction into Mercury.

153. The dissolution of gold must be perfected by nature and not by hands.

154. When gold is joined with its Mercury, then it is in the form of gold, but the most preparation is in the Calx.

155. There is a question amongst the Wise men, if the Mercury of Luna joined with the Mercury of gold may be had in the place of the philosophical Menstruum.

156. The Mercury of Luna keeps the nature of a man, and two men can generate no lesser than two women.

157. For the extracting of the Elixir you must get the most pure substance of Mercury.

158. He who will work, must work in the sublimation of the two luminaries.

159. Gold gives a golden and Silver gives a silvery Tincture, but he who knows (how) to tinge the Mercury with Silver or Gold has a great secret.

FINIS

A Word from the Publisher

Thank you for purchasing this small work from The R.A.M.S. Library of Alchemy. During his lifetime, Hans Nintzel was dedicated to the identification, acquisition, study, retyping and, when necessary, translation of what he considered to be the most important known works on Alchemy. Hans was assisted by his sparse network of fellow Alchemists, all members of the Restorers of Alchemical Manuscripts Society (R.A.M.S.). I was an active member of R.A.M.S.

My goal is to publish all of the works originally made available through R.A.M.S. as photocopies. To facilitate this, I have chosen to have the books professionally printed. I also have a few titles that I intend to add to the original R.A.M.S. Library, selected by strict criteria established by Hans.

The works from the original R.A.M.S. Library are republished by R.A.M.S. Publishing Company in the collection, "The R.A.M.S. Library of Alchemy," with permission of the Estate of Hans W. Nintzel.

If you have a work on Alchemy that you believe should be a part of the R.A.M.S. Library, please contact me through R.A.M.S. Publishing Company.

Philip N. Wheeler

www.ingramcontent.com/pod-product-compliance
Lightning Source LLC
Chambersburg PA
CBHW080809180526

45168CB00006B/2385